The Logica Yearbook

2010

The Logica Yearbook 2010

Edited by
Michal Peliš
and
Vít Punčochář

ISBN 978-1-84890-038-7

College Publications
Scientific Director: Dov Gabbay
Managing Director: Jane Spurr
Department of Computer Science
King's College London, Strand, London WC2R 2LS, UK

www.collegepublications.co.uk

Original cover design by Laraine Welch
Printed by Lightning Source, Milton Keynes, UK

Preface

The international symposium *Logica* has a long and rich tradition which goes back to the year 1987. *Logica* is organized by the Department of Logic of the Institute of Philosophy of the Academy of Sciences of the Czech Republic. The main purpose of these symposia is to provide a platform for the exchange of ideas among logicians from the whole world who can present the results of their research and discuss them with their colleagues. Every year about thirty papers are presented. Presentations are devoted to various branches of logic.

Participants of the symposia have the opportunity of publishing their contributions in *The Logica Yearbook* series. We have the pleasure of introducing the latest volume of the proceedings which contains most of the papers presented at *Logica 2010*, the 24th symposium in the series, that took place in the Franciscan monastery of Hejnice (Czech Republic) in June 2010.

The papers cover a wide variety of topics connected with logic. You can find papers concerning mathematical and philosophical logic, the history and philosophy of logic, and the logical analysis of natural language.

Both the *Logica* symposium and this book are the result of the joint effort of many people, who deserve our warmest thanks. We thank the Institute of Philosophy and especially its director, Pavel Baran, for his support. The conference was also made possible by the Grant Agency of the Czech Republic, which provided significant support by financing the grant project no. P401/10/1279. Special thanks also go to the Bernard Family Brewery of Humpolec, traditional sponsor of the social program of the symposium.

We are very grateful to the staff of Hejnice Monastery. It is with deep regret that we inform you that Miloš Raban, the director of Hejnice Monastery, has recently passed away.

We would also like to thank Karel Chvalovský for the layout of this volume and Petra Ivaničová who provided invaluable assistance to the organizers and took care of the participants at the conference. Many thanks go to College Publications and its managing director Jane Spurr. Last, but not least, we would like to thank all the authors for their exemplary collaboration during the editorial process.

Prague, May 2011   Michal Peliš and Vít Punčochář

i

Contents

Deflationism and Ontological Commitments

Theodora Achourioti*

Abstract

The deflationist argues that truth is not a substantial notion, primarily because adding a truth-predicate to a base theory like PA does not incur ontological commitments. Precisely what ontological commitments amount to, however, is a real issue. We first review how adding truth to PA by means of a satisfaction predicate leads to new properties and relations, and what the deflationist has to say thereof. We then offer an example of the addition of truth to a weaker arithmetical theory to argue that it clearly incurs ontological commitments. We argue further that even in the seemingly innocent case of PA as the base theory, the deflationist conclusion may be too strong. We suggest that the increase in the representational power of a theory that arises from the addition of a truth-predicate provides support for the substantial character of truth.

1 Introduction

The study of truth can be conducted in natural language, which already contains a truth-predicate, or in a formal language that does not already contain such a predicate, such as the language of Peano arithmetic.[1] In the formal literature on truth, a standard way (which

*This work is supported by the project 'The Origins of Truth and the Origins of the Sentence' funded by The Netherlands Organisation for Scientific Research, NWO.

[1]Following the standard literature PA is considered here as the base theory not containing a truth-predicate. However, strictly speaking, this is wrong, because PA defines partial truth-predicates in the following sense: given a class of formu-

originates in Tarski) of judging the substantiveness of truth in a precise manner is to look at the contribution of the truth-predicate when added to a theory that does not already contain it. Formal languages have an advantage over natural language in this respect, and one may hope that the study of truth in the formal setting will disclose something about the nature of truth as such.

Among the main tenets of deflationism is that the truth-predicate does not incur substantial content when added to a base theory. By substantial content the deflationist understands ontological commitments. This is a choice that the deflationist makes, for evident philosophical reasons: contrary to the substantial theorist, the deflationist thinks that it is not the role of truth to determine how things really are. Deflationism breaks tradition with the so-called substantial theories of truth, such as correspondence, coherence and pragmatic theories. These theories tend to either load truth with metaphysical weight or to reduce truth to other more basic notions; in both cases truth acquires substantial content. It is wrong, the deflationist argues, to burden truth with such high expectations.

That truth does not incur ontological commitments, is only one of the three main philosophical tenets of deflationism. The deflationist also thinks that the notion of truth is essentially captured by the T-schema, that is, an equivalence of the form $T(\ulcorner\phi\urcorner) \leftrightarrow \phi$. This schema has been taken to show the transparency and lightness of the notion of truth: Frege already observed that the truth-predicate is transparent in the sense that the meaning of the truth of a statement is no different from the one of the statement itself (Frege, 1918/1977). The T-schema is also taken to show the simplicity of rules that govern the truth-predicate: if ϕ holds, one can safely assert $T(\ulcorner\phi\urcorner)$, and vice versa.

But not all that the deflationist has to say about truth is negative. What the deflationist also emphasizes is that truth is not eliminable from the language (here meaning the natural language), as the re-

---

las of quantifier complexity n, it is possible to define in PA a truth-predicate of complexity $n+1$ which is defined for all formulas in PA of complexity n. Such results hold also for arithmetical theories weaker than PA, for example, the so-called bounded arithmetics. So it seems difficult to make sense of the notion of an arithmetical theory not containing some form of the truth-predicate. Therefore, in the present context, the phrase 'not containing a truth-predicate' should be understood as 'not containing a truth-predicate that is applicable to all formulas of PA'.

dundancy theorists have thought. Having the truth-predicate in a language allows the expression of statements that would not be expressible otherwise, namely, generalisations and blind ascriptions.[2] By quantifying over sentence positions, which is essentially what the truth-predicate allows to do, one can express a generalisation without having to enumerate its instances, and as a result infinite conjunctions/disjunctions can also be expressed.[3] Not having to be explicit about the sentence that is deemed to be true allows also the expression of blind ascriptions, since one can then quantify over a sentence that is denoted by means of a definite description.

In what follows we question the role that the deflationist assigns to ontological commitments in judging the substantiveness of truth, and by doing so we cast doubt on the doctrine that the power of the truth-predicate to express generalisations is all there is to the notion of truth. After an attempt to understand what ontological commitments really amount to, we review how adding truth by means of a satisfaction predicate to PA leads to new properties and relations, and what the deflationist has to say thereof. We then bring forward an example of adding truth to a weaker arithmetical theory to show that it clearly incurs ontological commitments. We argue further that even with the seemingly innocent case of PA as the base theory, the deflationist conclusion may be too strong. We suggest that the increase in the representational power of a theory that arises with the addition of a truth-predicate provides support for the substantial character of truth.

2 Deflationism and ontological commitments

2.1 Ontological commitments

In science, it is an important question whether a new theory enriches a certain ontology with new objects. Scientists are often driven to postulate new entities in order to explain properties of others. A question of direct verification of such entities then naturally follows, as this has been the case with such familiar physical entities as the

---

[2] Formally, the second is a special case of the first. See (Halbach, 2001).

[3] Without that implying that the infinite conjunctions/disjunctions and their expression by means of the truth-predicate have the same meaning. See (Gupta, 2001).

electron: Thomson postulated that the 'electron' was a new entity that explains charge, whereas the term was previously used to characterize a property found in existing objects. His experiments were aimed at showing that such an entity really exists.

This is admittedly a simplified picture of what goes on in science, however, it helps to convey the pre-theoretic intuition behind ontological commitments. Things are less straightforward in the context of formal (logical or arithmetical) theories, where it is not clear how to translate the situation just described. The main difference lies in that there is no independent means of verification of entities outside the context of the formal theory. This implies that the criteria for deciding whether something counts as an ontological commitment must in a sense be internal to the theory itself.

The clearest and standard definition of what 'ontological commitments' in a theory amount to one can find in Quine's slogan: 'To be is to be the value of a variable' (Quine, 1948/2003). With this formula Quine seeks to clarify, first, what one is *not* committed to. One is not committed to presupposing the existence of an entity when declaring of this entity that it does not exist: naming an entity does not force objective reference, for names can be transformed into predicates and definite descriptions (in accordance with Russell's theory of definite descriptions) which no object (i.e. nothing in the range of the bound variables) of the theory makes true. But this does not imply 'ontological immunity', Quine hastens to argue. What a theory *is* committed to are 'those and only those entities to which the bound variables of the theory must be capable of referring in order that the affirmations made in the theory be true' (Quine, 1948/2003, p. 15). Whether there are such entities depends on the ontology of one's theory; the formula does not decide this:

> ...how are we to adjudicate between rival ontologies? Certainly the answer is not provided by the semantical formula 'To be is to be the value of a variable'; this formula serves rather, conversely, in testing the conformity of a given remark or doctrine to a prior ontological standard. We look to bound variables in connection with ontology not in order to know what there is, but in order to know what a given remark or doctrine, ours or someone else's, says there is. (Quine, 1948/2003, p. 15)

As is evident from this passage, Quine's slogan—'to be is to be the value of a variable'—is elliptical as a definition of ontological commitments. The missing part is that something meets the criterion of being the value of a bound variable when it makes a formula true. It follows that the definition presupposes a particular notion of truth, the kind of truth that requires substitution of variables by objects, and an ontology that provides the range of the bound variables.[4]

The notion of conservativity has been considered extensively in the literature as a serious candidate for deciding the substantiality of truth.[5] If adding a truth-predicate leads to a non-conservative extension of the base theory, arithmetical facts expressible in the base theory can be proven that are not provable in this theory alone. One can then inquire whether these new commitments imply ontological commitments in the Quinean sense.

It has been pointed out that conservativity may be an unreasonable demand to impose on the deflationist, and that one should not expect importing truth into a theory to be an all-innocent act. In explaining the latter, Halbach (2001) distinguishes between conservativity over logic and conservativity over arithmetic, and with logic as the base theory, he shows that coding and identity are enough to yield a non-conservative extension with the addition of the truth-predicate. In the extended theory at least two different elements can be proven to exist as opposed to the weak ontological commitment of logic alone, which is a non-empty domain. Notice that this is fully in line with Quine's definition: the two different elements are objects that bounded variables of the existential formula 'there exist at least two elements' ($\exists x \exists y (x \neq y)$) range over. And concerning arithmetic, although a conservative extension is possible (e.g. the set of T sentences for arithmetical formulas), this theory of truth is not interesting enough even for the deflationist.

All this implies that it is within non-conservative extensions that the deflationist has to address the question of ontological commitments. With ontological commitments understood in the Quinean sense, the substantiality of truth becomes relative to the ontology of the chosen base theory and the particular axioms for truth adopted.

---

[4]This idea of a range of bound variables partially determines the properties of truth, since it allows classical negation. In intuitionism the space of constructions does not constitute a well-defined range; it is this which gives negation its non-classical properties.

[5]See, for example, (Shapiro, 1998).

The matter cannot even be decided for PA itself, because extending
PA with compositional axioms for truth while also allowing the truth-
predicate to occur in the induction schema is equivalent to a particular
theory which quantifies over sets of natural numbers. This equivalence then gives one the choice to either stay on the level of first-order
quantification, in which case bound variables still range over natural
numbers only, or move to the level of second-order quantification in
which case new ontological commitments are clearly enforced. This
last option is of course not the choice of the deflationist. But before
we proceed with the deflationist's choice, we include an interlude on
some basic facts about PA and truth.

2.2   The deflationist interpretation of basic facts about truth

With PA as the base theory, if $\phi(x)$ is a formula, the comprehension
axiom says:

(i) $\exists y \forall x \big(x \in y \leftrightarrow \phi(x) \big)$

This axiom is derivable, once one has a satisfaction predicate T
satisfying:[6]

(ii) $\forall x \big(T(\ulcorner \phi(u) \urcorner, x) \leftrightarrow \phi(x) \big)$

Comparison of (i) and (ii) shows that the satisfaction predicate T
is like $\in$, and by introducing $\exists$ we get:

(iii) $\exists y \forall x \big(T(y, x) \leftrightarrow \phi(x) \big)$

The sets of natural numbers introduced by the comprehension axiom are arithmetically definable and hence can be coded by a single
natural number. The comprehension axiom defines a set, the elements
of which are the natural numbers satisfying the arithmetical formula
and some of these numbers may themselves represent sets of natural
numbers[7] but they need not (not all natural numbers are codes). To

---

[6]The comprehension axiom is restricted by whatever restrictions one introduces in order to ensure a consistent addition of the satisfaction predicate. No
specification of such restrictions is made here, because we seek to make a general
point.

[7]For example, $\phi(x)$ iff x is the code of a formula that means divisible by a
prime number n. Take the code of this arithmetical formula and by means of the

return to Quine's definition, it is by construction that the variable x in (iii) still ranges over natural numbers, so at first sight the deflationist is not forced to accept new ontological commitments.

Due to the satisfaction predicate, familiar objects, i.e. natural numbers that act as codes of formulas, now acquire new properties and relations. They acquire new properties because they now denote sets and they inherit the properties of these sets. For example, some of these numbers code infinite sets, which sometimes represent real numbers.[8] They also acquire new relations. Two numbers $k > l$ may represent real numbers that stand in the opposite relation. Also, the numbers coded may have a dense ordering, which natural numbers do not have. These properties and relations are new: the ordering of the real numbers, for example, cannot be derived from the ordering of the natural numbers coding them.

The situation can now be described as follows:

> So one can take the fact that truth theories allow us to simulate fragments of analysis to show that truth theories introduce substantial new ontological commitments and are thus inflationary. But the message can also be taken to be that this fact shows that fragments of analysis can be interpreted as carrying no extra commitments over arithmetic and can thus be interpreted in a deflationary way. (Horsten, forthcoming)

The deflationist must deny that the properties and relations introduced by the addition of a satisfaction predicate entail new ontological commitments. These properties and relations are not available before the addition of the truth-predicate in the sense that they could not have been expressed in the base theory alone. But do they carry ontological weight?

It is worthwhile to pause once more and reflect on how these properties and relations come about. A natural number that does duty as the code of $\phi(x)$ takes on a new role thanks to the satisfaction predicate: whereas in the base theory it represents an arithmetical formula, it now represents a set, namely, the set of numbers that satisfy the formula. That the code already exists as an object in the base theory

---

satisfaction predicate you can then construct a set which has all these infinitely many sets as elements (i.e. all natural numbers divisible by 3, all natural numbers divisible by 5, all natural numbers divisible by 7 etc).

[8]Whether all infinite sets represent real numbers depends on the coding scheme.

is what ultimately allows us to introduce the ∃ and thereby generate the comprehension axiom: we quantify over an object in our base
theory.

The situation gives the deflationist two options. To the extent
that he takes the newly formed sets to carry ontological weight, (s)he
must see them as denoting something that already exists in the base
theory in some form. One can, for example, argue that expanding the
language with a truth-predicate has the power to unfold the reality
that is hidden in the base theory by allowing us to extract more information from the objects of the theory. This does not mean that
the information is—strictly speaking—new. It is ultimately because
of PA and not the satisfaction predicate that the classes of objects
defined by it are the ones that they are.

The other option for the deflationist is to admit that the truth-
predicate brings with it a radical shift, which, however, does not entail
ontological commitments. In this case the classes of objects definable
by means of the satisfaction predicate are better understood as 'virtual
classes' that do not have any substantial ontological weight. One can
argue that this is so because these sets enter the theory essentially in
virtue of their names, and having a name for a set does not argue for
its existence.[9] Names are not generally the case for sets: there are
uncountably many subsets of the natural numbers and only countably
many of those have names. For the infinite subsets of the natural
numbers that do not have names, the question of existence, if raised,
cannot be evaded. But a nominalist way out is possible otherwise,
even if it is not a decisive argument as such.

A similar line to take is to emphasize that the arithmetically definable properties in the base theory are ontologically prior to the sets
that they define by means of a truth-predicate. This is the proposal
made in the following passage:

> ... the truth-predicate introduces classes into our ontology: the
> truth-predicate allows us to inflate a no-class theory into a
> class theory. At least on the surface it looks as if the truth-
> predicate introduces a form of realism about classes. ... But
> it can also be maintained that the classes (properties) that are
> introduced by the truth-predicate are purely 'virtual'. Since
> we only quantify over classes (properties) that are defined by

---

[9]In Quine's terms, names do not commit to objective reference.

predicates, we can try to say with a straight face that we are
quantifying only over predicates. (Horsten, forthcoming)

3   More than meets the eye. . .

3.1   Ontological commitments in a narrow sense

PA has so far been taken as the base theory to which the satisfaction
predicate is added. This, however, may not be the right choice of a
base theory since PA is infinitistic and things look very different if one
starts from a finitistic theory such as PRA; finitistic proofs correspond
roughly to computations or combinatorial manipulations. We will see
that addition of the truth-predicate over this weak base theory (such
that the truth-predicate is allowed in the induction schema) is a very
substantive addition.

We will not discuss in detail arguments that the deflationist may
have for preferring PA over PRA as the base theory. It is our view
that there are various reasons why PA is not the right choice of a
base theory for what the deflationist wants to do, namely judge the
substantiveness of truth by adding it to a theory that is devoid of
truth. One reason for this is that in PA there are already partial
truth-predicates that can be defined; so it is not as if one adds a
truth-predicate to a truth-free theory. Another reason is that one can
judge the substantive contribution of truth only when it is added to
sufficiently weak theories, such as PRA.[10]

PRA has a quantifier-free induction axiom, or in other words,
an induction axiom for predicates only. But if one adds the truth-
predicate and allows it to occur as a predicate in the induction scheme,
then suddenly one has induction over all arithmetical formulas, i.e. the
induction scheme of PA. Roughly speaking, the truth-predicate lifts a
finitistic to an infinitistic theory, or to phrase it in Hilbert's terms, a
theory that deals with real entities to a theory that deals with ideal
entities, which, as we know, he would like to have seen eliminated.
Epistemological attitudes toward these two theories may differ vastly.

As explained earlier, one can talk of 'virtual classes' only because
one can pretend to operate with the name of the class, that is, the
Gödel number of the formula defining the class instead of the class itself. But this is not always possible, as one can see when considering

---

[10]In the end it is doubtful whether the experiment that the deflationist wants
to conduct can indeed be carried out in the context of arithmetic.

partial and total functions. Take, for example, the Ackermann function, formulated here in such a way that the definedness conditions are made explicit ($\downarrow$ indicates that the function is defined; because $\downarrow$ occurs only in antecedents the definition of the Ackermann function is still universal):

$$\forall m, n \Big[\big(m = 0 \rightarrow A(m, n) = n + 1\big)$$
$$\vee \big(m > 0 \wedge n = 0 \wedge A(m - 1, 1)\downarrow \rightarrow A(m, n) = A(m - 1, 1)\big)$$
$$\vee \big(m > 0 \wedge n > 0 \wedge A(m, n - 1)\downarrow \wedge \forall k A(m - 1, k)\downarrow$$
$$\rightarrow A(m, n) = A(m - 1, A(m, n - 1))\big) \Big]$$

One can give a name or code to the algorithm defining the function and this can be done in PRA already. In PRA, however, this function is not total, yet it can be proven to be total once one adds a truth-predicate and allows it in the induction scheme.[11]

Define the domain of the Ackermann function as follows: the pair $\langle m, n \rangle$ is in the domain if A applied to $\langle m, n \rangle$ yields a value. This is an arithmetical formula, which by the comprehension axiom corresponds to a set. Since PRA cannot prove the totality of the Ackermann function, there exists a model of PRA, in which the domain is equal to $\omega \times \omega$, but there also exists a model (namely, a model in which the function is not total) in which the domain is different from $\omega \times \omega$. In the standard model, one can prove that the function is total. But the trouble is that PRA has many models which are non-standard, and every model in which the Ackermann function is not total must be a non-standard model.

One clearly cannot talk here of the domain of this function as a 'virtual class' abstracting from ontological commitments. What this example shows is that a name does not necessarily have the same interpretation in every mathematical environment. Put differently, the name of a class does not determine its extension; the latter depends on the surrounding axioms. It follows that the deflationist has neither of the two choices outlined in the previous section. Neither can he take it that the satisfaction predicate helps to unveil the domain of the function that is somehow hidden in the base theory. Nor can he pretend

---

[11]The same can be achieved by the restricted second-order induction axiom together with the comprehension axiom.

that the new extension of the class is not of ontological status, given that (s)he follows Quine's definition of ontological commitments.

3.2   Ontological commitments in a broad sense

Examples like the one discussed in the previous section point to ontological commitments in a clear Quinean sense. But this is not the only way that the addition of the truth-predicate may be seen to incur ontological commitments. In this section we question the deflationist interpretation of seemingly innocent cases like the one of Section 2.2, where the option to dismiss ontological commitments seems unproblematic.

There is no doubt that 'commitment' is too strong a term to describe the effect of the truth-predicate when added to PA. But lack of clear ontological commitment also seems to be the wrong basis on which to draw the strong conclusion that truth is an insubstantial notion. Although it is unobjectionable that clear ontological commitments imply substantiality (as in the previous section), incurring no such commitment does not imply that the opposite is the case. One is not 'committed' to the ontology of the newly formed classes, but not in the same way that one is not committed, for example, to a specific coding scheme for coding arithmetical formulas. In the case of the coding scheme, talk of commitment is totally off the mark, since different codes will represent the same class when used to code the same arithmetical formula.[12] In the first case, however, commitments may eventually arise through the interaction of the truth theory with other theories. What 'virtual classes' add is an increase in representational capacity, through which new possibilities open up that were not previously available, such as the interaction of arithmetic with other domains; for instance, they allow the physicist to use the theory of differential equations due to the fact that real numbers can be represented.

Whether increased representational capacity in this case entails additional ontological commitments is not something for the truth theory to decide for itself. The normative force of a commitment comes from outside the truth theory; for instance, it comes from the physical world when the physicist postulates a new entity to explain

---

[12]Besides, a commitment to a coding scheme would be a commitment to a representation, and not an ontological commitment.

a certain property that (s)he has identified. The deflationist perspective will, of course, be that ontological additions do not come from the truth theory but the one it interacts with. However, it is the representational power of the truth-predicate that creates the conditions for ontological commitments to arise. A term from a different field may help here by way of analogy. In cognitive science, 'affordances' denote possibilities of action latent in the agent's environment, which the agent is or is not aware of. One can see the newly formed sets as affordances of PA in the environment of the satisfaction predicate and an appropriate coding scheme.

From this perspective, the 'representational power' of truth seems much more fitting than 'ontological commitments' for judging whether truth is a substantial notion or not. By 'representational power' we mean the capacity to represent something not previously accessible, regardless of whether this leads to a radical shift in ontology or not. In the end, Quine's definition is still lurking in the background; but now, the emphasis is on potential and not actual ontological commitments.

> I advanced an explicit standard whereby to decide what the ontological commitments of a theory are. But the question what ontology actually to adopt still stands open, and the obvious counsel is tolerance and an experimental spirit. (Quine, 1948/2003, p. 19)

All this says something about the normative orientation of the deflationist when it comes to interpreting facts concerning the formal features of truth, which is to explain away what truth adds to a base theory by showing that it does not incur new ontological commitments. But if this is all one is interested in, a great deal seems to be going missing, which may in fact speak in favor of the substantial character of truth. The message here is that much more can be gained by focusing on what truth adds to a theory rather than trying to explain it away.

4   Conclusion

We have discussed one of the main deflationist theses, which states that truth is an insubstantial notion because when added to a truth-free base theory it does not incur additional ontological commitments. We first required a clear definition of what counts as an ontological commitment, and we used Quine's standard definition to this end. Section 3.1 countered the deflationist thesis by illustrating that when added to weaker than PA arithmetical systems, truth clearly does imply ontological commitments in the Quinean sense. Section 3.2 further questioned the deflationist thesis, and with it the overall deflationist normative orientation, for even those cases that seem at first compatible with a deflationist interpretation, such as the addition of a satisfaction predicate to PA. We proposed that increased representational power provides support for the substantiveness of truth: it creates the conditions for potential ontological commitments that may arise through the interaction of the truth theory with another theory.

The above two objections make use of ontological commitments in a narrow and broad sense, corresponding to actual and potential ontological commitments respectively. The deflationist cannot ignore ontological commitments in the narrow sense, although (s)he may well think that ontological commitments in a broad sense do not constitute a refutation of the deflationist view. After all, (s)he does emphasize the expressive role of the truth-predicate. Even if this is the case, the deflationist will at least have to modify her/his claim that the expression of generalisations is all that is positive to say about truth.

References

Frege, G. (1918/1977). Thoughts. In P. Geach (Ed.), *Logical investigations*. New Haven: Yale University Press.

Gupta, A. (2001). A critique of deflationism. In M. P. Lynch (Ed.), *The nature of truth*. MIT.

Halbach, V. (2001). How innocent is deflationism? *Synthese*(126), 167–194.

Horsten, L. (forthcoming). *The Tarskian Turn: Deflationism and Axiomatic Truth*. MIT.

Quine, W. (1948/2003). On what there is. In *From a logical point of view* (2nd ed.). Harvard University Press.

Shapiro, S. (1998). Proof and truth: through thick and thin. *Journal of Philosophy*, *95*, 493–521.

Theodora Achourioti
ILLC/Philosophy Department, University of Amsterdam
Oude Turfmarkt 141—room 2.11
e-mail: t.achourioti@uva.nl

Grounding and Definiendum-Sensitivity

Emil Badici

Abstract

One way to avoid the semantic paradoxes is to incorporate in the definition of a truth-predicate the requirement that meaningful attributions of truth to a sentence must be grounded in non-semantic facts. The Liar sentence is the standard example of a groundless sentence, but many other sentences fail to meet the grounding requirement. The purpose of this paper is to show that a definition of a truth-predicate in terms of grounding might be definiendum-insensitive. Informally, a definition is definiendum-sensitive if it introduces a different meaning than some other definition which differs from it only in having a syntactically different definiendum. Definiendum-sensitivity is an undesirable feature if the definition is intended to capture an antecedently grasped concept.

One way to avoid the semantic paradoxes is to incorporate in the definition of a truth-predicate the requirement that meaningful attributions of truth to a sentence must be grounded in non-semantic facts. The Liar sentence is the standard example of a groundless sentence, but many other sentences fail to meet the grounding requirement. The purpose of this paper is to show that a definition of a truth-predicate in terms of grounding might be definiendum-sensitive. Informally, a definition is definiendum-sensitive if it introduces a different meaning than some other definition which differs from it only in having a syntactically different definiendum. Definiendum-sensitivity is an undesirable feature if the definition is intended to capture an antecedently grasped concept.

To get a preliminary idea of this phenomenon, consider an implicit definition of truth in terms of a T-schema restricted to exclude its contradictory instances but nothing more. Horwich (1990), for instance, suggests that the schema should be restricted as little as possible in order to make it consistent. The view has been attacked for many good reasons, including the charge that it is ad-hoc and that restricting the schema to a maximally consistent set of instances turns out to be impossible (McGee, 1992), but an additional objection drawing on the notion of definiendum-sensitivity remains significant for it can be generalized to other definitions as well. It turns out that two predicates, T_1 and T_2, defined by their corresponding restricted T-schemas would exclude different T-biconditionals. In particular, if L_1 is the Liar sentence defined below, then B_1 would have to be excluded (since it is contradictory) while B_2 would not.

(L_1) L_1 is not T_1.

(B_1) L_1 is T_1 if and only if L_1 is not T_1.

(B_2) L_1 is T_2 if and only if L_1 is not T_1.

Although this does not show that a definition of this sort would fail to provide its definiendum with meaning, it does however show that it would be definiendum-sensitive in a sense that might not have been intended. This idea lends itself to a more precise treatment when the notion of grounding is made to play a central role in the definition of truth.

1   Grounding and truth

The notion of grounding has a wide range of applications in both mathematics and semantics. The idea that truth, for instance, can be understood in terms of grounding is already a familiar theme. Herzberger (1970) draws an analogy between the semantic paradoxes and the paradoxes that occur in set theory, such as Mirmanoff's paradox of the groundless class, and proposes that a theory of semantics should adopt a restrictive rule, the semantic grounding condition, which would ban groundless sentences (in the sense that they do not make statements) in just the same way that regularity axioms ban groundless sets in set theory. Groundless sentences would include sentences which generate a vicious circle (for instance, the Liar sentence and the Truth-Teller) but also sentences which generate a vicious semantic regress. For instance, the infinite sequence of sentences S_n, where $n \geq 0$,

(S_n) S_{n+1} is true.

does not generate a vicious semantic circle, but a vicious semantic regress.

A more rigorous and technical definition of grounding for formal non-bivalent languages is provided by Kripke (1975). He shows how a first-order language which contains no semantic predicates and whose predicates are totally defined can be expanded by adding a new predicate whose extension is precisely the set of true sentences of the expanded language. I assume familiarity with Kripke's construction, so I will only mention some of those aspects of it which are relevant to the current project. The initial language, L, is an interpreted first-order language with connectives '$\sim$', '$\&$', '$\rightarrow$' and quantifiers '$\exists$' and '$\forall$' and a finite or denumerable list of primitive predicates. The domain D is nonempty and the n-ary predicates are interpreted as n-ary relations on D. L^+ is the language obtained by adding to L a new predicate, $T(x)$, which is characterized by an ordered pair (S_1, S_2) consisting of its extension and antiextension. At the initial stage of the construction S_1 and S_2 are both identical to the empty set, but they are expanded in a monotonic way at each successive stage until a fixed point is reached, where the extension of $T(x)$ coincides with the set of true sentences of the language. A sentence is called grounded if and only if it has a truth-value at the smallest fixed point and ungrounded otherwise.

The truth-predicate introduced in this way is merely partially defined: it has an extension and an antiextension but it is undefined for groundless sentences such as the Liar, the Truth-Teller and other paradoxical sentences. However, there is a heavy price that needs to be paid in order to ensure that the truth-predicate has the right extension. Although this account entails that the Liar is not true, this fact cannot be expressed in the language. In general, there appear to be two options for handling the truth-value gaps: interpret $T(\ulcorner A \urcorner)$ as a truth-value gap if and only if A is a truth-value gap or close-off the gap and interpret $T(\ulcorner A \urcorner)$ as false when A is either false or a truth-value gap. Kripke chooses the first option which, in his view, reflects an early stage in the acquisition of the language, but leaves the second option open as reflecting a later stage. The first option appears to be more appealing because both truths and falsities would be grounded in non-semantic facts. In the second case, on the other hand, the falsity of $T(\ulcorner A \urcorner)$ would be ultimately derived from a semantic fact: the fact that A is not assigned a truth-value at the minimal fixed point. This could perhaps be interpreted as a violation of the grounding requirement, but further reflection suggests that it is just an instance of another type of grounding. There is no good reason to deny that truths and untruths can be grounded in semantic facts.[1] Semantic facts do have grounding powers just like non-semantic facts. The central aspect of grounding is that the process of assigning a truth-value to a sentence is finite, regardless of whether it ends with semantic or non-semantic facts. The sentence ' 'Snow is white' is meaningful', for instance, is grounded in the semantic fact that 'Snow is white' does have a meaning. Similarly, the untruth of 'All mimsy were the borogoves', to use a standard example of a nonsensical sentence from Lewis Carroll, is grounded in the semantic fact that it is meaningless. The idea that truth should be grounded in non-semantic facts is the result of an oversimplification. For most purposes, the behavior of a truth-predicate is better studied on the model of a language which has a completely non-semantic vocabulary to which a new truth-predicate is added.

The more general question that arises now is what happens if the language already contains some partially defined predicates. It is unfortunate that Kripke's account, as well as other accounts inspired by it, are limited to languages whose predicates are fully defined and which have no truth-value gaps before the new predicate is added to the language. If one aims at providing a model that can be applied to the semantics of natural languages, one has to expand the account to cover languages which already contain gaps or ungrounded sentences and, more generally, to expandable language. Although Kripke claims that "it is unnecessary to suppose, as we have for simplicity, that all the predicates in L are totally defined" (Kripke, 1975, p. 720, fn. 20) he does not provide a sketch of how the account could be developed.

---

[1] Certainly, if semantic facts are ultimately reducible to non-semantic facts, then there would be no difference between the two types of grounding, but the accuracy of this reductive view of semantics is far from being obvious.

If a predicate, P, other than the truth-predicate, has application gaps, then this gappiness is presumably a semantic fact independent of the definition of the truth-predicate. Therefore, it is natural to think that the sentences that attribute the truth-predicate to $\ulcorner Px \urcorner$ or to $\ulcorner \sim Px \urcorner$, where x is one of the application gaps of P, are grounded and untrue.[2] The semantic fact that a given sentence of the original language is ungrounded grounds the fact that it belongs to the antiextension of the newly introduced truth-predicate.

The interesting case occurs when the language L is expanded with two truth-predicates defined by a Kripke-style construction, because the same Liar sentence would be a truth-value gap relative to one truth-predicate and part of the antiextension of the other. L_1, the Liar sentence corresponding to T_1, would be assigned neither to the extension nor to the antiextension of T_1 at the minimal fixed point, but it would be assigned already at the first stage to the antiextension of T_2 because its gappiness relative to T_1 can be established prior to the construction for T_2. Since T_1 and T_2 would fail to have the same meaning, the definition of truth would be in this case definiendum-sensitive.

2 Yablo on definitions and grounding

One can get a better insight into the role of grounding in the definition of a truth-predicate by reflecting on a more general theory of definitions. I will make use of the framework offered by Yablo (1993) to provide a general account of definitions in terms of grounding which encompasses both consistent and inconsistent definitions. Intuitively, grounding requires that the application of the newly defined predicate P to a certain object should be derivable from non-P facts or from facts which are themselves derivable from non-P facts, but there are different possible ways of spelling out this intuition. For Yablo, a starting point in understanding the connection between definitions and grounding is to think of the latter in terms of Tarski's recursively defined notion of satisfaction. Let $Px =_{\mathrm{def}} \varphi(x)$, be a definition of P. In order to determine whether an object satisfies the definiendum, one needs to make use of the recursive rules for simple predicates,

---

[2]This is by no means the only possible option, but it certainly captures an important aspect of truth.

negation, conjunction and quantifiers. The recursive clauses would suffice to determine whether the definiendum applies to an object as long as the definition is explicit, but there are other options available. Yablo distinguishes four types of definition: i) explicit (P does not occur in the definiens), ii) inductive (P occurs only positively in the definiens), iii) antiinductive (P occurs only negatively in the definiens) and iv) coinductive (P occurs both positively and negatively in the definiens).[3] The first two types of definitions are always consistent, but definitions of the last two types might be inconsistent. In the case of inductive definitions, the fact that the definiendum is true of a given object is sometimes derivable only from the antecedently established fact that it is true of other objects. Thus, Yablo points out, inductive definitions rely (in addition to the standard recursive rules which I will henceforth ignore) on the principle (ΔT) below which makes it possible for previous results to be re-entered in the iterative process.[4]

$$(\Delta T)\ T(\varphi, x) \Rightarrow T(P, x)$$

In addition to (ΔT), antiinductive definitions rely on the rule (ΔF),

$$(\Delta F)\ F(\varphi, x) \Rightarrow F(P, x)$$

because the fact that P is true of a certain object is sometimes derived from the fact that it is false of some other object. However, the simple rules, (ΔT) and (ΔF), do not suffice for coinductive definitions because they "offer no way of reflecting on the grounding process and incorporating the results of that reflection back into the process" (Yablo, 1993, p. 160). For this reason, (ΔF) is replaced by a more powerful reflective grounding rule, according to which "when you know enough to refute all possible proofs of x's claim to satisfy the definiens, you may conclude that the definiendum is false of x" (Yablo, 1993, p. 160). More precisely, the rule is (ΔR):

$$(\Delta R)\ \theta \Rightarrow F(P, x), \text{ where } \theta \text{ makes } x \text{ ungroundable.}$$

---

[3] See Yablo (1993, pp. 167–69) for a precise definition of positive and negative occurrences of a predicate in a formula.

[4] $T(\varphi, x)$ and $F(\varphi, x)$ abbreviate x satisfies φ and x does not satisfy φ, respectively.

A set of premises, θ, is said to make x ungroundable just in case $T(\varphi, x)$ cannot be derived even if θ is supplied with all $F(P, x)$ compatible with the current information.[5] Yablo's account of grounding is very significant in that it expands the domain of facts that could play a grounding role. Truths derived by appealing to (ΔR) are grounded in semantic facts.

3 Truth-definitions and definiendum-sensitivity

Kripke's definition of truth does not fit the pattern of definitions offered by Yablo. However, Yablo points out that the interpretation given to the truth-predicate when the minimal fixed point stage is reached is equivalent to the following simultaneous definition of truth and falsity:[6]

φ is true iff $\varphi = \ulcorner Ra \urcorner$ and a's referent belongs to R's extension

  or $\varphi = \ulcorner \sim\psi \urcorner$ and ψ is false

  or $\varphi = \ulcorner \psi \,\&\, \chi \urcorner$ and both ψ and χ are true

  or $\varphi = \ulcorner \forall x \psi(x) \urcorner$ and all its instances are true

  or $\varphi = \ulcorner \psi$ is true$\urcorner$ and ψ is true.

φ is false iff $\varphi = \ulcorner Ra \urcorner$ and a's referent belongs to R's antiextension

  or $\varphi = \ulcorner \sim\psi \urcorner$ and ψ is true

  or $\varphi = \ulcorner \psi \,\&\, \chi \urcorner$ and either ψ or χ are false

  or $\varphi = \ulcorner \forall x \psi(x) \urcorner$ and at least one of its instances is false

  or $\varphi = \ulcorner \psi$ is true$\urcorner$ and ψ is false.

In Yablo's view, the ordinary notion of truth is better captured by a slightly different simultaneous definition which replaces 'not true' for 'false' in the last line of the definition of falsity and otherwise remains

---

[5] One example of a coinductive definition (which turns out to be inconsistent) is, for Yablo, the definition of an anteger, where 'similar' means equidistant from zero:

  (A) x is an anteger $=_{\mathrm{def}} x$ is 0 or x is an integer similar to some non-anteger.

For an integer different from zero, reflective grounding entails that it is an anteger if and only if it is not an anteger.

[6] See (Yablo, 1993, p. 165).

the same. Kripke's is an inductive (and consistent) definition, because 'true' and 'false' occur only positively in the definiendum, while Yablo's modified definition is coinductive (and inconsistent). The difference is very important, because the latter definition introduces the process of reflective grounding. If none of the rules make it possible to prove that ψ is true, then, by reflective grounding, $\ulcorner \psi$ is true$\urcorner$ can be declared false.

Yablo's (ΔR) rule does not by itself generate definiendum-sensitivity because it is too restrictive. By always closing-off the gaps, it excludes the possibility of partially defined predicates. Moreover, there is a sense in which it is not the case that all possible proofs that x satisfies the definiens can be refuted before the reflection stage. In its original form, reflective grounding could occur before all the semantic information relevant to the predicate has been exhausted, because this very process of reflective grounding would constitute a new piece of semantic information. For this reason, (ΔR) would have to be replaced by some weaker reflective grounding rule. Although a weaker reflective grounding rule might not be available for all predicates, the following truth-specific reflective rule captures what is needed in the case of a truth-predicate:

> (ΔR_t) If one can prove that after all the semantic information relevant to P has been exhausted, it remains impossible to prove that either x satisfies or it does not satisfy $\varphi(x)$, then $F(T, \ulcorner Px \urcorner)$ and $F(T, \ulcorner \sim Px \urcorner)$.

This means that if the complete semantic information relevant to a predicate P leaves its status relative to x unspecified,[7] then the sentences that attribute the truth-predicate to $\ulcorner Px \urcorner$ or to its negation are false (' "Harry is bald" is true' is a false sentence if Harry is a borderline case). If the language contains two truth-predicates, T_1 and T_2, then the rule (ΔR_t) is the source of asymmetry. (ΔR_t) can be used to generate $F(T_1, \ulcorner \sim T_2 a_2 \urcorner)$, where a_2 is the Liar sentence relative to T_2, because the semantic information relevant to T_2 has been exhausted before the reflective stage, but not $F(T_2, \ulcorner \sim T_2 a_2 \urcorner)$, because in this case the semantic information relevant to T_2 has not

---

[7]The status of x remains unspecified when x is a gap of P, but also when the definition of P is inconsistent, as in some of the examples discussed by Yablo. For this reason, (ΔR_t) would be general enough to capture what is essential in both Kripke's consistent definition and in Yablo's inconsistent definition of truth.

been exhausted. If so, then the definition of the truth-predicate is definiendum-sensitive, because the two truth-predicates introduced in the same way are not synonymous. A Liar sentence would be a truth-value gap relative to its own truth-predicate, but it would belong to the antiextension of the other truth-predicate.

References

Herzberger, H. (1970). Paradoxes of Grounding in Semantics. *Journal of Philosophy*, *67*, 145–167.

Horwich, P. (1990). *Truth*. Oxford: Basil Blackwell.

Kripke, S. (1975). Outline of a Theory of Truth. *Journal of Philosophy*, *72*, 690–716.

McGee, V. (1992). Maximal Consistent Sets of Instances of Tarski's Schema (T). *Journal of Philosophical Logic*, *21*(3), 235–241.

Yablo, S. (1993). Definitions, Consistent and Inconsistent. *Philosophical Studies*, *72*, 147–175.

Emil Badici
Texas A&M University—Kingsville
MSC 182 / 1115 University Blvd
Kingsville, TX 78363
e-mail: kueb2005@tamuk.edu

A Semantics for Counterfactuals Based on Fuzzy Logic

Libor Běhounek[*]   Ondrej Majer[†]

Abstract

Lewis–Stalnaker's semantics for counterfactuals is based on the notion of similarity of possible worlds. Since the general notion of similarity is prominently studied in fuzzy mathematics, where it is modeled by fuzzy equivalence relations, it is natural to attempt at reconstructing Lewis' and Stalnaker's ideas in terms of fuzzy similarities. This paper sketches such a reconstruction; full details will be presented in an upcoming paper. We demonstrate that the approach is viable, adequate with respect to the expected properties of counterfactuals, and provides meaningful generalizations of the classical account.

1 Introduction

Counterfactual conditionals, or simply counterfactuals, are conditionals with false antecedents; i.e., conditionals of the form "if it were the case that A, then it would be the case that C", written $A \mathbin{\square\!\!\rightarrow} C$. Their logical analysis is notoriously problematic: if analyzed straightforwardly by material implication, they would always come out true; nevertheless, intuitively some counterfactuals seem to be true while others false. Consider, e.g., Goodman's (1947) example of the pair of counterfactuals regarding a piece of butter that has been consumed and was never heated:

---

[*]Supported by grants No. IAA900090703 of GA AV ČR and P202/10/1826 of GA ČR.

[†]Supported by grant No. ICC/08/E018 of GA ČR (a part of ESF EUROCORES–LogICCC project FP006).

(a) If that piece of butter had been heated to 150 F, it would have melted.

(b) If that piece of butter had been heated to 150 F, it would not have melted.

Clearly one of these two sentences (presumably the latter) should be considered false, even though both have false antecedents.

The logical analysis of counterfactual conditionals has for a long time been an important issue in philosophical logic. A notable early attempt to propose an adequate semantics for counterfactuals was done by Nelson Goodman (1947), who also stressed its importance, not only for logic, but as well for the philosophy of science. The most influential analysis was provided independently by Robert Stalnaker (1968) and David Lewis (1973). Their solution is based on the notion of similarity which compares possible worlds according to their (subjective) closeness to the actual world. In the present paper, we shall mostly follow Lewis' formulation of the semantics. According to Lewis, a counterfactual $A \mathbin{\Box\!\!\rightarrow} C$ is true in the actual world iff either A does not hold in any world (then $A \mathbin{\Box\!\!\rightarrow} C$ is trivially true), or all worlds in which both A and C hold are closer (in a given ordering of worlds by similarity to the actual world) to the actual world than some world(s) in which A and $\neg C$ hold.

The general notion of similarity (not just of worlds) is prominently studied in fuzzy mathematics, where it is modeled by fuzzy equivalence relations (Zadeh, 1971). It is, therefore, natural to attempt a reconstruction of Lewis' and Stalnaker's ideas in terms of fuzzy similarities. This paper offers such a reconstruction in the framework of t-norm fuzzy logics (see esp. Hájek, 1998). We demonstrate that the approach is viable and that the resulting semantics is adequate with respect to the expected properties of counterfactuals (i.e., it validates their intuitive and invalidates their counter-intuitive properties). The merits of the fuzzy approach to counterfactuals are briefly discussed in the concluding section. The paper provides a sketch of the approach; more details will be found in an upcoming full paper.

2 Lewis' semantics of counterfactuals

Lewis' approach is based on the notion of (subjective) *similarity* of possible worlds.[1] The intuitive notion of similarity of worlds can be formally rendered in a number of ways, and the rendering impacts the resulting properties of the logic of counterfactuals. Lewis himself (1973) provides a variety of counterfactual logics differing in the properties of the similarity ordering, and also several reformulations that replace the similarity order by alternative notions. Here we shall present his formulation in terms of the relation of *closeness* between possible worlds, with notation $v \leq_w v'$ understood as "the world v is at least as close to the actual world w as the world v'".

Various sets of natural properties of the closeness relation can be assumed, generating different logics of counterfactuals. For instance, the following assumptions yield Lewis' counterfactual logic VC:

- Strict minimality of the actual world: $w <_w v$ for any $v \neq w$

- Linearity: for any v, v', either $v \leq_w v'$ or $v' \leq_w v$

- Transitivity: if $v \leq_w v'$ and $v' \leq_w u$, then $v \leq_w u$

The counterfactual conditional $A \mathbin{\Box\!\!\rightarrow} C$ is defined to be true in a world w with respect to a closeness relation $\leq_w$, iff:

- Either for all v, $v \not\models A$ (i.e., there are no A-worlds), or

- There is a v' such that $v' \models A \wedge C$ and for every v, if $v \models A \wedge \neg C$ then $v' <_w v$ (i.e., there is an AC-world which is closer to the actual world then any $A\neg C$-world)

Lewis (1973) showed that the logic VC can be axiomatized by the

---

[1] The similarity relation is a primitive parameter of the semantic model. How the relation is actually obtained is of no concern for the logic of counterfactuals: it is simply assumed to be given, similarly as subjective probabilities are assumed to be given in probability theory.

following axioms:

$$A \mathbin{\square\!\!\rightarrow} A \tag{1}$$
$$(\neg A \mathbin{\square\!\!\rightarrow} A) \rightarrow (A \mathbin{\square\!\!\rightarrow} B) \tag{2}$$
$$(A \mathbin{\square\!\!\rightarrow} \neg B) \vee (((A \wedge B) \mathbin{\square\!\!\rightarrow} C) \leftrightarrow (A \mathbin{\square\!\!\rightarrow} (B \rightarrow C))) \tag{3}$$
$$(A \mathbin{\square\!\!\rightarrow} B) \rightarrow (A \rightarrow B) \tag{4}$$
$$(A \wedge B) \rightarrow (A \mathbin{\square\!\!\rightarrow} B) \tag{5}$$

plus the axioms of classical propositional calculus and the rules of modus ponens, substitution, and the rule of deduction within conditionals,

$$\frac{(B_1 \wedge \ldots \wedge B_n) \rightarrow C}{(A \mathbin{\square\!\!\rightarrow} B_1) \wedge \ldots \wedge (A \mathbin{\square\!\!\rightarrow} B_n) \rightarrow (A \mathbin{\square\!\!\rightarrow} C)} \tag{6}$$

The modality $\square$ of necessity, with the usual meaning of truth in all accessible worlds, is definable in Lewis's system VC in terms of counterfactual implication:

$$\square A \equiv \neg A \mathbin{\square\!\!\rightarrow} \bot \tag{7}$$

Various intuitively plausible properties of counterfactuals are derivable in the logic VC, for instance the following ones:

$$\square(A \rightarrow B) \rightarrow (A \mathbin{\square\!\!\rightarrow} B) \tag{8}$$
$$\square \neg A \rightarrow (A \mathbin{\square\!\!\rightarrow} B) \tag{9}$$
$$\square B \rightarrow (A \mathbin{\square\!\!\rightarrow} B) \tag{10}$$
$$(A \mathbin{\square\!\!\rightarrow} B) \mathbin{\&} \square(B \rightarrow B') \rightarrow (A \mathbin{\square\!\!\rightarrow} B') \tag{11}$$
$$(A \mathbin{\square\!\!\rightarrow} B) \mathbin{\&} \square(A \leftrightarrow A') \rightarrow (A' \mathbin{\square\!\!\rightarrow} B) \tag{12}$$

By (4) and (8), the logical strength of counterfactual implication is intermediate between those of material and strict implication. Unlike material conditionals, counterfactual conditionals do *not* obey, i.a., the following laws:

- Weakening: $\dfrac{A \mathbin{\square\!\!\rightarrow} C}{(A \mathbin{\&} B) \mathbin{\square\!\!\rightarrow} C}$

- Contraposition: $\dfrac{A \mathbin{\square\!\!\rightarrow} C}{\neg C \mathbin{\square\!\!\rightarrow} \neg A}$

- Transitivity: $\dfrac{A \mathbin{\Box\!\!\rightarrow} B,\, B \mathbin{\Box\!\!\rightarrow} C}{A \mathbin{\Box\!\!\rightarrow} C}$

Informal counterexamples to these rules as well as formal counterexamples in Lewis' semantics can easily be constructed (or see Lewis, 1973).

3 Formal fuzzy logic

Fuzzy logics are many-valued logics suitable mainly for *gradual* predicates, such as *tall, old, warm,* etc., whose underlying quantities can be measured by real numbers (e.g., height in cm's, age in years, temperature in Fahrenheits, etc.). In order to be measured on a common scale, the underlying quantities are conventionally transformed into the unit interval $[0, 1]$ (or another suitable algebra); the transformed values are called the *degrees* of the gradual predicates (e.g., the degrees of tallness, warmness, etc.). Certain algebraic operations are defined on $[0, 1]$ that represent logical connectives and quantifiers, and a degree-based consequence relation between $[0, 1]$-valued gradual propositions is studied.

Different fuzzy logics arise by different admissible choices of algebraic operations that realize logical connectives on $[0, 1]$. A prototypical (and probably the best known) example of fuzzy logic is infinite-valued Łukasiewicz logic; other members of the family are, for instance, Gödel–Dummett logic, product fuzzy logic, fuzzy logics BL, MTL, etc. For the sake of simplicity, we shall in this paper restrict our attention to Łukasiewicz logic, even though all our considerations, derivations, and semantic examples actually work for any well-behaved formal fuzzy logic.[2] For basic information on formal fuzzy logic see, e.g., (Hájek, 2010); more technical details can be found, e.g., in (Hájek, 1998). In this section, we shall briefly recall basic definitions of Łukasiewicz logic.

In the standard $[0, 1]$-valued semantics of Łukasiewicz logic, propo-

---

[2]Namely, for any extension of the fuzzy logic MTL, which is arguably the weakest t-norm–based fuzzy logic suitable for fuzzy mathematics.

sitional connectives are realized by the following operations on $[0,1]$:

$$x \to y = \max(1 - x + y, 1)$$
$$x \mathbin{\&} y = \min(0, x + y - 1)$$
$$x \wedge y = \min(x, y)$$
$$x \vee y = \max(x, y)$$
$$x \leftrightarrow y = 1 - |x - y|$$
$$\neg x = 1 - x$$
$$\Delta x = 1 \text{ iff } x = 1, \text{ otherwise } \Delta x = 0$$

Notice that Łukasiewicz logic possesses two conjunctive connectives: idempotent $\wedge$ and non-idempotent $\&$. The presence of these two conjunctions is not surprising, considering the fact that Łukasiewicz logic belongs to contraction-free substructural logics (Ono, 2003) all of which possess two conjunctions.[3] The connective Δ, expressing the full degree of a gradual proposition, enables interpretation of classical logic within Łukasiewicz logic (as its connectives and quantifiers behave classically on the values 0 and 1).

Tautologies of propositional Łukasiewicz logic are those formulae that always evaluate to 1. The set of propositional tautologies of Łukasiewicz logic is finitely axiomatizable by a Hilbert-style calculus (see Hájek, 1998).

First-order Łukasiewicz logic extends the propositional syntax in the usual way. Semantically, n-ary first-order predicates are interpreted by n-ary functions from a fixed set of individuals (the universe of discourse) to $[0,1]$. The quantifiers $\forall$ and $\exists$ are interpreted, respectively, as the infimum and supremum of the degrees of all instances of the quantified formula. First-order Łukasiewicz logic can be axiomatically approximated by adding Rasiowa's (1974) axioms for quantifiers to propositional Łukasiewicz logic.[4]

---

[3] Łukasiewicz logic, similarly as other contraction-free substructural logics, also has another, non-idempotent disjunction $x \oplus y = \min(1, x + y)$, which, however, will not be needed in this paper. (In (29) below, which is the only place where disjunction occurs in this paper, the variants with $\oplus$ and $\vee$ happen to be equivalent, since one of the disjuncts is bivalent.)

[4] Rasiowa's axioms are only complete with respect to a generalized semantics of first-order Łukasiewicz logic, which evaluates formulae in more general algebras than $[0,1]$. The standard $[0,1]$-valued first-order semantics is not finitarily axiomatizable, though it can be axiomatized by an infinitary rule—see (Hay, 1963).

4   Higher-order fuzzy logic

First-order Łukasiewicz logic is sufficiently strong for supporting non-trivial axiomatic mathematical theories. For smooth mathematical work it is expedient to define a Russell-style simple type theory over Łukasiewicz logic (or Henkin-style higher-order Łukasiewicz logic, see Běhounek & Cintula, 2005). The latter is an axiomatic theory over multi-sorted first-order Łukasiewicz logic, with sorts of variables for individuals from a fixed universe of discourse (zeroth-order variables) and for predicates of all orders $n \geq 1$ and arities $k \geq 0$ (n-th-order variables). The language of the theory comprises:

- The primitive memberships predicates $\in$ between successive sorts,

- The bivalent identity predicate $=$ on each sort,

- The functions $\langle \ldots \rangle$ for tuples of each order and arity, and

- The comprehension terms $\{x \mid \varphi\}$ of order $n + 1$ for each well-typed formula φ and the variable x of each order n.

The theory is axiomatized by the following axiom schemata:

- The identity axioms $x = x$ and $x = y \rightarrow (\varphi(x) \rightarrow \varphi(y))$, for each well-typed formula φ and each sort of variables;

- The usual technical axioms for handling tuples of each arity and order;

- Comprehension axioms $y \in \{x \mid \varphi(x)\} \leftrightarrow \varphi(y)$ for each well-typed formula φ and each sort of variables;

- Extensionality axioms $(\forall x)\Delta(x \in A \leftrightarrow x \in B) \rightarrow A = B$ for all orders.

In the intended models of the theory,[5] individual variables range over a fixed set X (the universe of discourse). First-order k-ary pred-

---

Nevertheless, Rasiowa's axiomatic approximation is sufficient for almost all practical purposes.

[5] Just like in classical higher-order logic, the theory of intended models is not recursively axiomatizable (as it interprets true arithmetic). The above axiomatization is sound with respect to intended models, but complete only with respect to a more general class of models. Nevertheless, the axiomatic approximation is again sufficient for almost all practical purposes.

icates are interpreted as $[0,1]$-valued fuzzy sets (for $k = 1$) or k-ary fuzzy relations (for $k > 1$) on X; i.e., as functions $X^k \to [0,1]$. Second-order k-ary predicates are interpreted as fuzzy sets (or relations) of fuzzy sets (or relations) of individuals; i.e., as functions $(X_1)^k \to [0,1]$, where X_1 is the range of 1st-order predicate variables. In general, $(n+1)$-st-order k-ary predicates are interpreted as fuzzy sets (or relations) of order $n+1$; i.e., as functions $(X_n)^k \to [0,1]$, where X_n is the range of n-th-order predicate variables. The membership predications $x \in A$ are assigned the value (in $[0,1]$) that the function interpreting A assigns to the interpretation of x.

Comprehension terms $\{x \mid \varphi(x)\}$ (of any order) denote fuzzy sets to which each element x belongs to the degree of $\varphi(x)$. Classical (bivalent) sets can be represented by fuzzy sets whose membership functions only take values in the two-element set of degrees $\{0,1\}$. The condition of being bivalent is expressible as $(\forall x)\Delta(x \in A \vee x \notin A)$, where $x \notin A$ abbreviates $\neg(x \in A)$.

Various fuzzy mathematical notions can be defined over Łukasiewicz logic by reinterpreting the formulae of classical mathematics in higher-order Łukasiewicz logic. In what follows, we shall need the following elementary defined notions:

$$\emptyset = \{x \mid 0\} \tag{13}$$
$$A \subseteq B \equiv (\forall x)(x \in A \to x \in B) \tag{14}$$

Moreover we shall use the following defined connectives that compare the degrees of formulae:

$$\varphi \le \psi \equiv \Delta(\varphi \to \psi) \tag{15}$$
$$\varphi = \psi \equiv \Delta(\varphi \leftrightarrow \psi) \tag{16}$$

The definition is justified by the fact that in Łukasiewicz logic (as well as in other fuzzy logics), the degree of $\varphi \to \psi$ is 1 iff the degree of ψ is at least as large as the degree of φ, and the degree of $\varphi \leftrightarrow \psi$ is 1 iff the degrees of φ and ψ are equal.[6]

---

[6]The defined connective $=$ of (16), expressing the identity of degrees, should not be confused with the identity predicates on each sort of variables: though denoted here by the same sign, they are always disambiguated by the context, as the arguments of the latter are terms, while the arguments of the former are formulae.

5 Similarity relations

Similarity relations are in fuzzy mathematics standardly modeled as fuzzy equivalence relations (Zadeh, 1971), i.e., binary fuzzy relations S that satisfy (to degree 1) the axioms of fuzzy reflexivity, symmetry, and transitivity in Łukasiewicz (or another fuzzy) logic:

$$\Delta(\forall x)Sxx \tag{17}$$

$$\Delta(\forall xy)(Sxy \rightarrow Syx) \tag{18}$$

$$\Delta(\forall xyz)(Sxy \mathbin{\&} Syz \rightarrow Sxz) \tag{19}$$

A stronger notion of *separated* (or *unimodal*) fuzzy equivalence (sometimes also called *fuzzy equality*) strengthens the reflexivity condition (17) by the additional requirement that only identical elements are fully similar:

$$(\forall xy)(\Delta Sxy \leftrightarrow x = y) \tag{20}$$

Let a fuzzy relation $S^M \colon X^2 \rightarrow [0,1]$ be the interpretation of the predicate S in a given standard $[0,1]$-valued semantic model M (with the universe X) for Łukasiewicz logic. Then the function $d(x,y) = 1 - S^M xy$ is a (bounded) pseudometric iff S is a fuzzy equivalence relation (i.e., satisfies (17)–(19) to degree 1) in M, and is a bounded metric iff the fuzzy equivalence is separated (i.e., satisfies (18)–(20) to degree 1 in M).[7]

The correspondence to (pseudo)metrics and the fact that they can be conveniently handled by means of formal fuzzy logic (cf. Běhounek, Bodenhofer, & Cintula, 2008) makes fuzzy equivalence relations a suitable representation of gradual relations of closeness or similarity. Consequently, they have become the standard model of similarity relations in fuzzy mathematics, and we shall call fuzzy equivalence relations simply (fuzzy) similarities further on. It should, however, be stressed that formal fuzzy logics admit other algebras of degrees besides the $[0,1]$ interval (see, e.g., Hájek, 1998). Fuzzy equivalence relations thus generalize real-valued (pseudo)metrics by admitting various scales of

---

[7]In a similar manner, fuzzy equivalence relations correspond to (generally unbounded) pseudometrics $d(x,y) = -\log S^M xy$ in product logic, pseudoultrametrics in Gödel logic, and various generalizations of pseudometrics in other fuzzy logics (differing in the operation used in the triangle inequality), and to corresponding kinds of metrics if the equivalence relation is separated.

abstract degrees of closeness (or similarity); consequently, modeling similarity or closeness by equivalence relations over fuzzy logic does *not* enforce measuring them by real numbers.

6   Similarity of worlds and degrees

Applying the apparatus of fuzzy similarities to Lewis' concept of the similarity of possible worlds, we assume that there is a fuzzy relation Σ on the (classical, bivalent) set W of possible worlds, which satisfies the axioms of fuzzy similarity (17)–(19). The formula Σxy is informally interpreted as "the world x is similar to the world y", and the degree (from $[0, 1]$ or another algebra of degrees admissible for Łukasiewicz logic) that is assigned to it in a particular model of (higher-order) Łukasiewicz logic is interpreted as the degree of similarity between these worlds.

The first idea how to model Lewis' ternary relation "the world x is at least as close to the actual world w as the world y" in terms of the similarity relation Σ is to define it straightforwardly as $\Sigma yw \leq \Sigma xw$, where $\leq$ is the degree-comparing connective (15). However, for reasons both methodological and technical, it turns out to be more appropriate to use a *fuzzy* comparison $\lesssim$ of degrees, rather than the bivalent comparison connective $\leq$, defining:

$$x \preccurlyeq_w y \ \equiv \ \Sigma yw \lesssim \Sigma xw. \tag{21}$$

The intended definition could be informally interpreted as "x is more or *roughly as* similar to w as y". This form follows the fuzzy paradigm to employ, whenever reasonable, a (fuzzy) indistinguishability (or similarity) relation rather than bivalent equality. The particular motivation for using $\lesssim$ instead of $\leq$ is the natural assumption that worlds indistinguishable from x (as regards their closeness to the actual world) should in the evaluation of counterfactuals play a rôle similar to that of x.

In order to define the fuzzy ordering $\lesssim$ of degrees (of similarity to the actual world) informally interpretable as "more or roughly as", we need to define another similarity, $\sim$, this time on degrees. Besides the usual axioms of similarity, we shall require $\sim$ to satisfy two additional conditions that specify the kind of similarity relations suitable for our purposes. Omitting the initial Δ's (for satisfaction to degree 1) and quantification over all degrees, the conditions on $\sim$ can be listed as follows:

$$(\alpha \sim \alpha) \tag{22}$$
$$(\alpha \sim \beta) \rightarrow (\beta \sim \alpha) \tag{23}$$
$$(\alpha \sim \beta) \& (\beta \sim \gamma) \rightarrow (\alpha \sim \gamma) \tag{24}$$
$$(\alpha \sim \beta) \rightarrow (\alpha \leftrightarrow \beta) \tag{25}$$
$$(\exists \beta \neq \alpha)(\beta \sim \alpha) \tag{26}$$
$$(\alpha \leq \beta \leq \gamma) \rightarrow ((\alpha \sim \gamma) \rightarrow (\alpha \sim \beta)) \tag{27}$$
$$(\gamma \leq \beta \leq \alpha) \rightarrow ((\alpha \sim \gamma) \rightarrow (\alpha \sim \beta)) \tag{28}$$

The conditions (22)–(24) are just the axioms of fuzzy similarity. The condition (25) of congruence with respect to the equivalence connective expresses the substitutivity of similar degrees in gradual inference: its equivalent formulation is $\alpha \& (\alpha \sim \beta) \rightarrow \beta$. It furthermore ensures that the similarity $\sim$ is separated, since $\Delta(\alpha \sim \beta) \leftrightarrow (\alpha = \beta)$ is a corollary of (25); in other words, each degree is *fully* similar only to itself. The condition (26) ensures that $\sim$ differs from the bivalent identity $=$, but still to each degree there are arbitrarily similar degrees.[8] The latter two conditions also ensure that $\sim$ is indeed fuzzy, as by (25) and (26) the relation $\sim$ has to be infinite-valued. Finally, the conditions (27)–(28) express the compatibility of $\sim$ with the ordering $\leq$ of degrees: it ensures that closer (in the sense of $\leq$) degrees are (non-strictly) more similar to each other.

By means of the relation $\sim$, the fuzzy comparison "more or roughly as" $\lesssim$ of degrees can be defined as follows:

$$\alpha \lesssim \beta \equiv (\alpha < \beta) \vee (\alpha \sim \beta). \tag{29}$$

The conditions (22)–(26) then entail the following corresponding prop-

---

[8]By the semantics of $\exists$ in Łukasiewicz logic, (26) ensures that for each α, the supremum of the *degrees of similarity* of other degrees to α is 1.

erties of $\precsim$:

$$(\alpha \sim \beta) \to (\alpha \precsim \beta) \tag{30}$$

$$(\alpha \precsim \beta) \,\&\, (\beta \precsim \alpha) \to (\beta \sim \alpha) \tag{31}$$

$$(\alpha \precsim \beta) \,\&\, (\beta \precsim \gamma) \to (\alpha \precsim \gamma) \tag{32}$$

$$(\alpha \precsim \beta) \to (\alpha \to \beta) \tag{33}$$

$$(\alpha \leq \beta) \to (\alpha \precsim \beta) \tag{34}$$

$$(\exists \beta \neq \alpha)(\beta \precsim \alpha) \tag{35}$$

$$(\alpha \leq \beta \leq \gamma) \to ((\gamma \precsim \alpha) \to (\beta \precsim \alpha)) \tag{36}$$

$$(\gamma \leq \beta \leq \alpha) \to ((\alpha \precsim \gamma) \to (\alpha \precsim \beta)) \tag{37}$$

The properties (30)–(32) state that $\precsim$ is a *similarity-based fuzzy ordering* (for which see Bodenhofer, 2000) with respect to the similarity $\sim$. The conditions (33)–(35) entail corollaries analogous to those of (25) and (26), namely:

$$\alpha \,\&\, (\alpha \precsim \beta) \to \beta$$

$$\Delta(\alpha \precsim \beta) \leftrightarrow (\alpha \leq \beta), \tag{38}$$

and also enforce $\precsim$ to be a fuzzy (rather than bivalent) ordering of degrees, strictly weaker than $\leq$. The sets of axioms (22)–(28) and (30)–(37) are in fact equivalent, as the former can be obtained from the latter if $\sim$ is defined as the symmetrization of $\precsim$, i.e., $\alpha \sim \beta \equiv (\alpha \precsim \beta) \wedge (\beta \precsim \alpha)$, or equivalently, $(\alpha \precsim \beta) \,\&\, (\beta \precsim \alpha)$.

By means of the fuzzy ordering $\precsim$ on degrees (of similarity of worlds), we can now define Lewis' ternary relation $\preccurlyeq$ of closeness of worlds as intended above in (21), namely:

$$x \preccurlyeq_w y \equiv \Sigma yw \precsim \Sigma xw,$$

interpreted as "x is more or roughly as similar to w as y".

7   Fuzzy semantics of counterfactuals

Having interpreted the similarity of worlds and Lewis' ternary relation of closeness of worlds in the fuzzy setting, we can now define the semantics of counterfactuals based on these notions. The primitive parameters of the semantics, supposed just to be given (cf. footnote 1 above), are the fuzzy similarity relation Σ on the (bivalent) set W of possible worlds and the fuzzy similarity relation $\sim$ on degrees (of similarity of possible worlds), assumed to satisfy the axioms given in Section 6. As usual in intensional semantics, the meanings of propositions are identified with (here in general fuzzy) sets of possible worlds.

The definition of the semantics of counterfactuals in the fuzzy setting will be based on the straightforward idea that the counterfactual $A \mathbin{\square\!\!\rightarrow} C$ is true in the world w if the closest (with respect to w) A-worlds are also C-worlds.[9] The A-worlds closest to w can be defined as the fuzzy set of *minima* of a fuzzy set A of worlds in a fuzzy relation (not necessarily an ordering) $\preccurlyeq_w$ (cf. Běhounek et al., 2008):

$$\mathrm{Min}_{\preccurlyeq_w} A \;=\; \{x \mid x \in A \wedge (\forall a)(a \in A \to x \preccurlyeq_w a)\},$$

i.e., the fuzzy set of elements of A at least as close to w as any element of A (which is the classical definition just reinterpreted in terms of Łukasiewicz logic). Basic properties of minima in fuzzy relations are easily derivable in higher-order fuzzy logic (cf. Běhounek et al., 2008).

Given a model M of higher-order Łukasiewicz logic and the parameters Σ and $\sim$, the semantic value (or *extension*) of the counterfactual $A \mathbin{\square\!\!\rightarrow} C$ in the possible world w can now be defined as follows:

$$(A \mathbin{\square\!\!\rightarrow} C)_w \;\equiv\; (\mathrm{Min}_{\preccurlyeq_w} A) \subseteq C, \tag{39}$$

where $\subseteq$ is fuzzy inclusion defined in (14). The definition expresses, in the fuzzy sense, that all $\preccurlyeq_w$-minimal (i.e., closest to w) A-worlds are also C-worlds. Notice that since the right-hand side of (39) is evaluated in Łukasiewicz logic, $(A \mathbin{\square\!\!\rightarrow} C)_w$ denotes a *degree* in $[0,1]$ (or another algebra admissible for Łukasiewicz logic).

The meaning (or *intension*) of the counterfactual proposition $A \mathbin{\square\!\!\rightarrow} C$ is then identified with the fuzzy set of possible worlds to which a possible world w belongs to degree $(A \mathbin{\square\!\!\rightarrow} C)_w$, i.e.,

$$(A \mathbin{\square\!\!\rightarrow} C) \;=\; \{w \mid (A \mathbin{\square\!\!\rightarrow} C)_w\}.$$

With these definitions at hand, a standard account of fuzzy intensional semantics for propositional logic of counterfactuals can already

---

[9]This actually corresponds more to Stalnaker's approach than to Lewis'. It will be seen in Section 8 (cf. footnote 10) that unlike in Stalnaker's bivalent framework, in our fuzzy setting this *does not* involve the implausible Limit Assumption.

be given. We omit it here for space restrictions; it runs along the usual lines, defining a formula A of propositional logic of counterfactuals to have degree 1 in a model (formed of the bivalent set W of worlds, the parameters Σ and $\sim$, and an evaluation of propositional variables in each world) if its intension in the model is the whole set W of its possible worlds, and to be a *tautology* (written $\models A$) if it has degree 1 in all models. A detailed account of the semantics as well as a thorough discussion of all defined notions will be given in a full paper (in preparation).

8   Properties of fuzzy counterfactuals

Various properties of fuzzy counterfactuals can be derived in higher-order Łukasiewicz logic. We give just a sample of our results here, omitting the proofs (which will be given in a full paper under preparation).

First, the condition (26) ensures that the fuzzy set of minima of A in $\preccurlyeq_w$ is non-empty iff A is non-empty. Consequently, in any world w in any given model, $(A \,\square\!\!\rightarrow C)_w$ has the full degree 1 for all C if and only if $A = \emptyset$. In other words, a counterfactual is trivially true only if its antecedent is impossible.[10]

Second, the undesirable properties of counterfactuals mentioned in Section 2 are indeed refutable in our framework: fuzzy models can be constructed (by adapting Lewis' bivalent counterexamples) that invalidate the rules of weakening, contraposition, and transitivity for counterfactual implication. The counterexamples disprove not only their graded variants, but also the weaker forms with premises satisfied to the full degree:

$$\not\models \Delta(A \,\square\!\!\rightarrow C) \to ((A \,\&\, B) \,\square\!\!\rightarrow C)$$

$$\not\models \Delta(A \,\square\!\!\rightarrow C) \to (\neg C \,\square\!\!\rightarrow \neg A)$$

$$\not\models \Delta(A \,\square\!\!\rightarrow B) \,\&\, \Delta(B \,\square\!\!\rightarrow C) \to (A \,\square\!\!\rightarrow C)$$

---

[10]Our definition (39), though straightforwardly using the fuzzy *minimum* of A in the closeness relation, thus need not use the implausible Limit Assumption (cf. Lewis, 1973, p. 20) that there always exist the closest among A-worlds (for $A \neq \emptyset$). In the fuzzy setting, the rather natural non-triviality condition (26) on the similarity of degrees automatically entails that the *fuzzy* set of minima of a non-empty fuzzy set in the fuzzy ordering by closeness is non-empty (though the minimal worlds need not exist to the full degree).

Third, the desirable properties of counterfactuals are valid in our framework. For instance, it is provable in higher-order Łukasiewicz logic that the strength of the counterfactual conditional is intermediate between the material and strict conditionals (cf. (4) and (8) above in Section 2):

$$\models \Box(A \to C) \to (A \,\Box\!\!\to C)$$
$$\models (A \,\Box\!\!\to C) \to (A \to C),$$

if the (S5-)modality of necessity is defined in the natural way as the proposition

$$(\Box A) \ = \ \{w \mid (\forall v)(v \in A)\}.$$

Many further classical tautologies involving counterfactuals are provable in higher-order Łukasiewicz logic, for instance the properties (1) and (9)–(12) as well as the rule (6) of deduction within conditionals. Some classical tautologies, however, only hold for full degrees: for instance, it is provable that

$$\models \ \Delta(\Box\neg A) \leftrightarrow \Delta(A \,\Box\!\!\to \bot), \tag{40}$$

but counterexamples can be given to this formula without the Δ's (cf. (7)); similarly for Lewis' axiom (2).

9   Conclusions

We have shown that basic ideas of Lewis–Stalnaker semantics of counterfactuals can be reconstructed in higher-order Łukasiewicz logic (in fact, mutatis mutandis, in every t-norm fuzzy logic). The semantics has been shown to be adequate to the intuitive understanding of counterfactual conditionals; i.e., it validates plausible properties of counterfactuals and invalidates implausible ones (at least of those considered in this paper).

The new formalization of Lewis–Stalnaker semantics is based on an application of a well-developed general theory of fuzzy similarity relations to the particular case of the similarity of possible worlds. Fuzzy logic, which can be interpreted as logic of gradual predications, is particularly suitable for the purpose, since the relation of similarity is indisputably gradual (some things are more similar than others). The ensuing fuzzy semantics moreover automatically accommodates counterfactuals that involve gradual predicates (i.e., such that come in degrees). It also admits the gradualness of counterfactuals themselves, i.e., the fact that some counterfactuals are perceived to be truer than others. The price paid for these advantages is the use of non-classical logic in the semantics; however, since t-norm logics are well-established (being not too different from intuitionistic or linear logic) and mathematics based on these logics is sufficiently developed (see Běhounek & Cintula, 2005; Běhounek et al., 2008), the cost is not too high and seems to be worth the gains.

The present paper only dealt with the fuzzy semantics of the logic of counterfactuals; its axiomatization as well as the study of its syntactic or semantic variants is left for future work.

References

Běhounek, L., Bodenhofer, U., & Cintula, P. (2008). Relations in Fuzzy Class Theory: Initial steps. *Fuzzy Sets and Systems*, *159*, 1729–1772.

Běhounek, L., & Cintula, P. (2005). Fuzzy class theory. *Fuzzy Sets and Systems*, *154*, 34–55.

Bodenhofer, U. (2000). A similarity-based generalization of fuzzy orderings preserving the classical axioms. *International Journal of Uncertainty, Fuzziness and Knowledge-Based Systems*, *8*, 593–610.

Goodman, N. (1947). The problem of counterfactual conditionals. *The Journal of Philosophy*, *44*, 113–128.

Hájek, P. (1998). *Metamathematics of fuzzy logic*. Dordercht: Kluwer.

Hájek, P. (2010). Fuzzy logic. In E. N. Zalta (Ed.), *The Stanford encyclopedia of philosophy*. World Wide Web. (Fall 2010 edition.)

Hay, L. S. (1963). Axiomatization of the infinite-valued predicate calculus. *Journal of Symbolic Logic*, *28*, 77–86.

Lewis, D. (1973). *Counterfactuals*. Oxford: Blackwell.

Ono, H. (2003). Substructural logics and residuated lattices—an introduction. In V. F. Hendricks & J. Malinowski (Eds.), *50 years of Studia Logica* (pp. 193–228). Dordrecht: Kluwer.

Rasiowa, H. (1974). *An algebraic approach to non-classical logics*. Amsterdam: North-Holland.

Stalnaker, R. (1968). A theory of conditionals. In N. Rescher (Ed.),

Studies in logical theory (pp. 98–112). Oxford: Oxford University Press.

Zadeh, L. A. (1971). Similarity relations and fuzzy orderings. *Information Sciences*, *3*, 177–200.

Libor Běhounek
Institute of Computer Science, Academy of Sciences of the Czech Rep.
Pod Vodárenskou věží 2, 182 07 Prague 8, Czech Republic
e-mail: behounek@cs.cas.cz
URL: http://www.cs.cas.cz/behounek

Ondrej Majer
Institute of Philosophy, Academy of Sciences of the Czech Rep.
Jilská 1, 110 00 Prague 1, Czech Republic
e-mail: majer@site.cas.cz

Considerations contra Cantorianism

Frode Bjørdal

Abstract

We present an introduction to librationism, which is a foundational system related to set theoretic foundations. It resolves paradoxes in a novel manner without compromising classical logical theorems even though it e.g. has as a theorem that Russell's sort (set) R of sorts not members of themselves is a member of itself and it also has as a theorem that R is not a member of itself. Librationism is semi-formal and negation complete. The semantics is based upon a semi inductive Herzbergerian process. We present many of the salient principles of librationism, and also show how the system avoids Cantor's conclusion that there are over-denumerable infinities by instead deducing the paradoxicality of an invoked sort. Surrounding issues including some concerning mathematical strength are also discussed.

With the avoidance of Russell's paradox and its cognates as one paramount motivation, and the avoidance of ungrounded mathematical objects as another, the twentieth century from early on saw the initiation of various foundational theories which altogether avoided an invocation of infinite power sets. This is famously the case in the predicativist tradition going back to Herman Weyl's *Das Kontinuum*, and further investigated later principally by Solomon Feferman, but also by others. This was clearly also an important motivational aspect of the perhaps less rigorously formulated original intent of Luitzen Brouwer's intuitionist program, and it is presently manifest much more precisely within parts of the intuitionist tradition in that Per Martin Löf's constructive Type Theory lacks an analogue of the infinitary power-set operation, as does Peter Aczel's constructive set theory CZF.

The reverse mathematics program initiated by Harvey Friedman seemed to have established that only a very small fragment of Second order Arithmetic suffices for ordinary classical mathematics, i.e. mathematics which is not concerned with purely set-theoretic issues or those of its "higher reaches". The system we propose, *(minimalistic) librationism*[1], has mathematical strength beyond the fully impredicative system Π_1^1–CA_0 plus the Bar-rule in a sense to be made more precise below. As librationism also sheds important light on other important phenomena, as the paradoxes, it seems that it may advantageously serve as a foundation for the mathematics needed for science, semantics and philosophical speculation.

I point out first in this separate paragraph that the last sentence in the previous paragraph may seem dated on account of other important very recent work by Friedman which shows that certain sentences of quite concrete mathematics in his Boolean Relation Theory needs something like ZFC plus the existence of all inaccessible strong Mahlo cardinals of finite order to be settled correctly. The interested reader should consult his home page for the book draft on Boolean Relation Theory. But Friedman has there also shown that what suffices to prove *the exotic cases* is ACA′ (which is often formulated as ACA_0 + the existence of the Turing jump for all subsets of ω) plus 1-Con(SMAH). The latter is an *arithmetical* schema which to the effect states the 1-consistency of Friedman's SMAH, i.e. that all Σ_1^0 sentences provable in ZF+$\bigwedge_{n=0}^{n=\infty}$ (there is a strong n-Mahlo cardinal) are also true. As Π_1^1–CA_0 plus the Bar-rule much exceeds ACA′, which is even weaker than ACA (=ACA(0) plus full induction), and our *semi formal* system librationism, for reasons that become clear below, settles all arithmetical sentences, we know *a priori* at a meta level that Friedman's exotic cases have librationist resolutions. How such resolutions may come about, and what they amount to, will remain to be seen. To find the resolutions will most likely not be a trivial exercise, perhaps quite on the contrary; it must, as is the standard procedure in isolating *partial* axiomatic and inferential principles of librationism, take its recourse

---

[1]I have coined the term "librationism" from the term "libration" which is used for certain oscillating phenomena e.g. in astronomy. This seems a useful name for our system as it is not taken, it reminds of the peculiar shift in perspectives which are involved in the theory's treatment of paradoxical phenomena and is also close in spelling and pronunciation to the term "liberalism" which I have used in some earlier lectures and publications. The latter term suggested itself because of the emancipatory feature that all set terms are allowed and dealt with in a comprehensive and, it is hoped, justified and edifying manner.

to the semi inductive semantics which we describe below. The point here is that we, from the librationist outlook, ultimately do not need to buy into ZFC-like points of view, or, more generally, opinions that include infinite power sets, in order to account even for Friedman's new exotic incompleteness phenomena.

There are two important traditions which we shall recall briefly in order to situate librationism. (1) Since the work of Saul Kripke and others there has been an explosive interest in self-referential truth. An important strand in that development was initiated independently by Hans Herzberger and Anil Gupta, and now serves as a background for semi inductive and revisionary style semantics. For us, the system LES introduced at the end of §69 in Andrea Cantini's important monograph has been influential. LES is an axiomatic theory of truth and abstraction justified by a semi inductive type semantics, and which respects classical logic. Librationism properly extends LES, and also respects classical logic. (2) Starting with Stanislaw Jaskowski's non-adjunctive system and the work of Newton da Costa, many formal calculi weaker than classical logic have been proposed in the paraconsistent tradition so as to allow naïve comprehension. But it cannot be seen that these views offer a satisfactory analysis of how mathematical objects are engendered by the accompanying nave comprehension principle. Often e.g. ZFC is merely taken for granted as an inside paradox free theory, and paradoxical phenomena are just assumed to be inside an outside shell. One may challenge that these views thus ride piggy-back on classical views without offering the analysis we are in need of and which should be of our interest. Still, as the acute reader will see, librationism shares some important features with paraconsistent approaches.

Let us for this exposition simplify and take our language to have parentheses for punctuation, infinitely many variables v_0, v_1, v_2, ... , connectives $\wedge$, $\vee$, $\equiv$, $\supset$, $\sim$, quantifiers $\forall$, $\exists$, epsilon $\in$, the truth operator T and the set builder $\{:\}$. Formation rules and inter-definability relations are as would be expected by my audience, with the addition that A is a formula only if $\mathsf{T}A$ is a formula. I point out that set brackets are not eliminable as in extensional set theories. Semantically, we rely on a semi-inductive process on ordinals. In our context we need no "boot-strapping policy". $\{v_i : A\}$ is a set-constant if A has at most v_i free. Fix an enumeration $e(0)$, $e(1)$, ... of all set-constants. $[A]$ is the Gödel-number of the formula A for a given coding. X is a function from ordinals to subsets of natural numbers and $\models$ a relation between such subsets and formulas as given by the double recursion: For any ordinal α,

(1) $X(\alpha) = \{ [A] : \exists \beta (\beta \prec \alpha \,\&\, \forall \gamma (\beta \preceq \gamma \prec \alpha \Rightarrow X(\gamma) \models A)) \}$

(2) $X(\alpha) \models \mathsf{T}A$ iff $[A] \in X(\alpha)$

(3) $X(\alpha) \models a \in \{ v_i : A \}$ iff $X(\alpha) \models \mathsf{T}A(a/v_i)$

(4) $X(\alpha) \models A \wedge B$ iff $X(\alpha) \models A$ *and* $X(\alpha) \models B$

(5) $X(\alpha) \models {\sim} A$ iff *not* $X(\alpha) \models A$

(6) If $a = e(i)$ then $X(\alpha) \models A(v_i)$ iff $X(\alpha) \models A(a)$

(7) $X(\alpha) \models \forall v_i A$ iff for all variables (names!) v_j, $X(\alpha) \models A(v_j/v_i)$.

By adapting results going back to Herzberger there will be a stabilization ordinal κ so that $X(\kappa) \models \mathsf{T}A$ iff $\forall \gamma (\gamma \succeq \kappa \Rightarrow X(\gamma) \models A)$. Notice that $X(\kappa) \models A$ or $X(\kappa) \models {\sim} A$, and also that by (7) the isolated *minimalist*[2] model, and thence the librationist system, is closed under the non-constructive Z-rule: If $X(\kappa) \models A(u)$ for all variables (or for all terms) u, then $X(\kappa) \models \forall x A(x)$. We at this point make a **crucial** shift in metalogical attention to $\{ A : X(\kappa) \models {\sim}\mathsf{T}{\sim}A \}$ as our designated model (modulo the enumeration e invoked in the semantical set up), and define:

$$ \Vdash A =_{\mathrm{D}} X(\kappa) \models {\sim}\mathsf{T}{\sim}A. $$

By induced principles at $X(\kappa)$ we have $X(\kappa) \models \mathsf{T}A$ iff *not* $X(\kappa) \models {\sim}\mathsf{T}A$, so we define A to be a *maxim*, $\Vdash_{\mathrm{M}} A$, iff $\Vdash A$ and *not* $\Vdash {\sim}A$. *Minors* are given by $\Vdash_{\mathrm{m}} A$ iff $\Vdash A$ and $\Vdash {\sim}A$. $r \in r$ for $r = \{ x : x \notin x \}$ is e.g. a minor. Theorems of classical logic are examples of maxims. It can be shown that identity can be introduced à la Leibniz. Induced inferential principles are unfamiliar, and one should not expect that these can be effectively circumscribed. It is noteworthy that the traditional inferential schema modus ponens holds for $\Vdash_{\mathrm{M}}$ (so we call this

---

[2] We think of the model and resulting librationist system as *minimalist* because the truth operator T has the empty extension at ordinal zero, i.e. $X(\emptyset) \models {\sim}\mathsf{T}A$ for all formulas A.

adjusted inferential schema *modus maximus* in the librationist framework) but not for $\Vdash_m$, as we do e.g. **not** have that the conjunction of $\Vdash_m A$ and $\Vdash_m A \supset \bot$ entails that $\Vdash_m \bot$. In consequence, librationism may be understood as a non-adjunctive system. All induced principles for truth and set-theoretic abstraction are, or so I argue, incontrovertible. One should, I again emphasize, appreciate that librationism is a semi-formal system, or framework, so that we will have that $\Vdash A$ or $\Vdash \sim A$ and that it is also closed under the Z-rule described above.

For the reader's benefit, I list some axiomatic and inferential principles of librationism in order to have a partial description (it is understood that all generalizations of instances of the following schemas are axioms, so that generalization is not a primitive inference rule):

L1M $\ A \supset (B \supset A)$

L2M $\ (A \supset (B \supset C)) \supset ((A \supset B) \supset (A \supset C))$

L3M $\ (\sim B \supset \sim A) \supset (A \supset B)$

L4M $\ A \supset \forall x A$, provided x is not free in A.

L5M $\ \forall x(A \supset B) \supset (\forall x A \supset \forall x B)$

L6M $\ \forall x A \supset A(t/x)$, if t is substitutable for x in A.

LO1M $\ \mathsf{T}(A \supset B) \supset (\mathsf{T}A \supset \mathsf{T}B)$

LO2M $\ \mathsf{T}A \supset \sim\mathsf{T}\sim A$

LO3M $\ \mathsf{T}B \vee \mathsf{T}\sim B \vee (\mathsf{T}\sim\mathsf{T}\sim A \supset \mathsf{T}A)$

LO4M $\ \mathsf{T}B \vee \mathsf{T}\sim B \vee (\mathsf{T}A \supset \mathsf{T}\mathsf{T}A)$

LO5M $\ \mathsf{T}(\mathsf{T}A \supset A) \supset (\mathsf{T}A \vee \mathsf{T}\sim A)$

LO6M $\ \exists x \mathsf{T}A \supset \mathsf{T}\exists x A$

LO7M $\ \mathsf{T}\forall x A \supset \forall x \mathsf{T}A$

LO8m $\ \mathsf{T}A \supset A$

LO9m $\ A \supset \mathsf{T}A$

LO10m $\forall x TA \supset T\forall x A$

LO11m $T\exists x A \supset \exists x TA$

Schemas that are maxims have a capital "M", while minor schemas, i.e. those that have minor instances, have a minuscule "m" at the end of its appellation as given here.

We next point out the librationist comprehension principle:

LCM $\forall x (x \in \{y : A\} \equiv TA(x/y))$, if x is substitutable for y in A.

We proceed to provide some salient inferential principles that we can show hold in librationism:

IR1: If $\Vdash_M A$ and $\Vdash_M A \supset B$ then $\Vdash_M B$ ("modus maximus")

IR2: If $\Vdash_m A$ and $\Vdash_M A \supset B$ then $\Vdash B$ ("modus subiunctio")

IR3: If $\Vdash_M A$ and $\Vdash_m A \supset B$ then $\Vdash_m B$ ("modus antecedentiae")

IR4: If $\Vdash_M A$ then $\Vdash_M TA$ ("modus ascent maximus")

IR5: If $\Vdash_m A$ then $\Vdash_m TA$ ("modus ascent minor")

IR6: If $\Vdash_M TA$ then $\Vdash_M A$ ("modus descent maximus")

IR7: If $\Vdash_m TA$ then $\Vdash_m A$ ("modus descent minor")

IR8: If $\Vdash_M {\sim}T{\sim}A$ then $\Vdash_M TA$ ("modus scandent maximus")

IR9: If $\Vdash_m {\sim}T{\sim}A$ then $\Vdash_m TA$ ("modus scandent minor")

IR10: If $\Vdash_M \forall x TA$ then $\Vdash_M T\forall x A$ ("modus Barcan")

IR11: If $\Vdash T\exists x A$ then $\Vdash \exists x TA$ ("modus attestor")

(Also: If $\Vdash_m T\exists x A$ then $\Vdash_m \exists x TA$)

IR12: If $\Vdash_m A$ and $\Vdash_m B$ then $\Vdash_m {\sim}T{\sim}A \wedge {\sim}T{\sim}B$ ("modus minor")

In this short presentation, the semantical verification of the axiomatic and inferential principles is left as an exercise to the reader. I also leave it as a problem to show that modus Barcan and modus attestor cannot be strengthened as one should intuitively expect.

We can by means of a fixed point construction going back to Andrea Cantini and Albert Visser isolate what we call *manifestation points*: If $A(x, y)$ is a formula with the free variables shown we can find a term h^A such that $\Vdash_M \forall z(z \in h^A \equiv \mathsf{T}\mathsf{T}A(z, h^A))$. To prove this, take ordered pairs, e.g. à la Kuratowski, and suppose $d = \{ \langle x, g \rangle : A(x, \{ u : \langle u, g \rangle \in g \}) \}$ and $h^A = \{ x : \langle x, d \rangle \in d \}$. Define $\mathrm{KIND}(x) =_D \forall y(\mathsf{T}y \in x \lor \mathsf{T}y \notin x)$. Taking H as the manifestation point of $\mathrm{KIND}(x) \land x \subset y$, we have that $\Vdash_M a \in H$ iff $\Vdash_M \mathrm{KIND}(a) \land a \subset H$. We can then show that we for $\mathbb{N} = \omega = \{ x : \forall y(\emptyset \in y \land \forall z(z \in y \supset z' \in y) \supset x \in y) \}$, with $z' = \{ w : w = z \lor w \in z \}$, have that $\Vdash_M \mathbb{N} \in H$. From what we pointed out in the second paragraph, we indeed have *fabulously* much more as regards maxims relative to H. We may intuitively think of H as the set of hereditarily non-paradoxical and, on account of our minimalist policy, wellfounded sets.

Let $\Vdash_M \mathrm{KIND}(f)$ and f a surjection from $\mathbb{N}$ to $V = \{ x : x = x \}$, and consider Cantor's $s = \{ x : x \in \mathbb{N} \land x \notin f(x) \}$, i.e. $s = \{ x : x \in \mathbb{N} \land \exists y(\langle x, y \rangle \in f \land x \notin y) \}$. Since $\Vdash_M \mathrm{KIND}(f)$ and $\mathrm{function}(f)$ and $\Vdash_M 8 \in \mathbb{N}$ this reduces to $\Vdash_M 8 \in s \equiv \mathsf{T}(8 \notin s)$. In the non-adjunctive framework of librationism it turns out that we only have the schemas $\Vdash_m A \supset \mathsf{T}A$ and $\Vdash_m \mathsf{T}A \supset A$ in full generality, and not always the conjunction of instances as minor schemas such as these have minor instances. This and further schemas and inferential principles only license the conclusion that $\Vdash 8 \in s$ and $\Vdash 8 \notin s$, i.e. $\Vdash_m 8 \in s$. So s turns out to be paradoxical, just as Russell's set. We underline that <u>Cantor's arguments are perfectly valid</u>, but in librationism we find that the appropriate assumption to be discarded in the reductio is the camouflaged assumption that s is non-paradoxical. Since f was taken as a KIND surjection from $\mathbb{N}$ to V, it is a fortiori onto the power set of $\mathbb{N}$.

Other Cantorian arguments for higher infinities are dislodged for quite analogous or parallel reasons. I here leave residual matters for the reader's meditation. It turns out as a consequence that the set of real numbers, taken e.g. by Dedekind cuts, is paradoxical, and so not *listable*. By this we mean that there is no non-paradoxical function from $\mathbb{N}$ which has exactly non-paradoxical real numbers as values; there are, I mention *en passant*, paradoxical Dedekind cuts or real numbers as e.g. $\{ x \in \mathbb{Q} : (x <^{\mathbb{Q}} 0 \wedge r \in r) \vee (x <^{\mathbb{Q}} 1 \wedge r \in r) \}$ where $\mathbb{Q}$ is the set of rational numbers, $<^{\mathbb{Q}}$ its standard ordering and r is Russell's paradoxical set $\{ x : x \notin x \}$. Taking the real numbers by e.g. Cauchy-sequences does of course not alter the essential dialectics of the situation. Our observations are such that we are justified in thinking that librationism *allows*[3] a non-paradoxical function from $\mathbb{N}$ onto V, and so also onto the set of real numbers. There are thus according to our librationist point of view no more real numbers than there are natural numbers.

It is noteworthy in all of this to realize that virtually all power sets are paradoxical in librationism. I leave this as an exercise. Hint: Consider the manifestation point $\mathring{a}$ such that $\Vdash_M \forall z (z \in \mathring{a} \equiv \mathsf{TT} z \notin \mathring{a})$ and evaluate $\mathring{a}$'s membership with respect to the power sets of any set m such that it is not a maxim that m is coextensional with $V = \{ x : x = x \}$.

We first pointed out that infinite power sets are not needed in order for the foundation of ordinary mathematics. Our results show that even if they are accommodated as in librationism, they do not here support Cantor's conclusion as paradoxical phenomena are accounted for in such a way as to forestall one of Cantor's essential though camouflaged assumptions. In this, we take librationism to *somehow* confirm the predicativist suspicions towards the use of power sets.

Cantorianism, as I understand it here, includes the point of view that there is at least one infinite cardinality larger than $\aleph_0$. The foregoing serves to challenge this tenet of Cantorianism and math-

---

[3] We have of course not by this shown that librationism as so far developed *has* such a surjection from $\mathbb{N}$ to V. I have shown in work that goes beyond what is presented here that we may enlarge the librationist language with a new set constant $\in$ and have its denotatum serve as a bijection from $\mathbb{N}$ to V by just slightly altering the semantical set up. Furthermore, and importantly, we may extend the librationist language with set constants $\in$ *and* T and have the denotatum of the former serve as a bijection as pointed out while the denotatum of the latter serves as a truth *predicate* on Gödel numbers of formulas; the semantical alterations needed are minor. The accommodation of a truth predicate depends upon having the universe denumerable. It is of strong philosophical interest to include a truth predicate because it helps us provide what I think is a favorable librationist account of the Liar's paradox and related self referential semantical phenomena. Mathematically, a (non-paradoxical) bijection $\in$ is very useful as it provides important choice principles in many settings where desirable.

ematical orthodoxy. There are further philosophical considerations which pull in the same direction that I in conclusion of this note just briefly mention: (1) Ontological economy suggests that Cantorianism is ontologically extravagant. (2) Cantorianism makes it impossible to presuppose an overarching philosophical meta-language. (3) If Cantorianism were true mathematical reality would need to exhibit disturbing essential traits of metaphysical incompleteness. (4) If we presuppose Cantorianism the question "How many objects are there?" becomes meaningless and cannot be answered; but it seems perfectly meaningful and in librationism has the answer that there are precisely countably infinitely many objects.

References

Bjørdal, F. (2005). There are Only Countably Many Objects. In *The LOGICA Yearbook 2004* (pp. 47–58). Prague: FILOSOFIA.

Bjørdal, F. (2006). Minimalistic Liberalism—an Adequate, Acceptable, Consistent and Contradictory Foundation. In *The LOGICA Yearbook 2005* (pp. 39–50). Prague: FILOSOFIA.

Cantini, A. (1996). *Logical Frameworks for Truth and Abstraction.* Elsevier.

Feferman, S. (1984). Toward Useful Type-Free Theories. *Journal of Symbolic Logic, 49,* 75–111.

Friedman, H. (n.d.). *Boolean Relation Theory and Incompleteness.* (downloadable from http://www.math.ohio-state.edu/ ~friedman/manuscripts.html)

Gupta, A. (1982). Truth and Paradox. *Journal of Philosophical Logic, XI*(1), 1–60.

Herzberger, H. (1982). Notes on Naive Semantics. *Journal of Philosophical Logic, XI*(1), 61–102.

Kripke, S. (1975). Outline of a Theory of Truth. *Journal of Philosophy, 72,* 690–716.

Frode Bjørdal
The Department of Philosophy, Classics & History of Art and Ideas
The University of Oslo
P.O.Box 1020, 0315 Blindern
e-mail: frode.bjordal@ifikk.uio.no
URL: http://www.hf.uio.no/ifikk/personer/vit/fbjordal/

Can Priest's Dialetheism Avoid Trivialism?

Massimiliano Carrara   Silvia Gaio

Enrico Martino

Abstract

Priest's *dialetheism* is the view that some sentences, called *dialetheias*, are both true and false. A crucial problem of dialetheism is that of avoiding *trivialism*, i.e. the consequence that *all* sentences are *dialetheias*. In the present paper we want to discuss Priest's strategy for avoiding *Curry's paradox*, from which trivialism follows. Besides, we will formulate a new version of Curry's paradox, using a notion of *naïve deducibility*, which, in our opinion, should be accepted by Priest himself.

1   Introduction

As is well known, dialetheism maintains the thesis that there are true contradictions, i.e. true sentences of form $(A \wedge \neg A)$, called *dialetheias*.[1] More generally, we will call dialetheia any sentence that is both true and false. In a reach series of papers and books (see, for example, Priest, 1979, 2002a, 2002b, 2006a, 2006b), Priest claims that dialetheism supplies the best solution to *the strengthen liar paradox*, a paradox originated from the following sentence:

(a):   (a) is not true

by holding that (a) is both true and not true. More generally, he holds that paradoxical self-reference sentences are dialetheias.

Priest's dialetheism has been extensively criticized in the literature (for an overview of criticism see Berto, 2007, part IV). In this paper we will not discuss the crucial problem concerning the acceptance of a dialetheia. Rather, we will focus on the following claims by Priest:

---

[1]G. Priest uses the terms 'dialetheias' and 'true contradictions' to indicate 'gluts', which in turn is a term coined by K. Fine in (Fine, 1975).

1. The presence of dialetheias does not entail trivialism:

   - a contradictory theory may be not explosive, i.e. it does not necessarily prove all sentences;

   - in a model containing a dialetheia not all sentences are necessarily true.

2. The meaning of logical constants is the same in the object language and in the metalanguage.

Of course, claim 1 is of vital importance for dialetheism: if a dialetheia implied everything, dialetheic theories would be of no interest.

Claim 2 should guarantee that dialetheias are not produced by altering *ad hoc* the intended meaning of logical constants. Using Priest's words (Priest, 2006b, p. 98):

> The distinction between a theory (...) and its metatheory makes perfectly good sense to a dialetheist. But there is no reason to insist that the metatheory must be stronger than, and therefore different from, the theory. The same logic must be used in both 'object language' and 'metatheory'.

In classical logic, explosion is produced using the rule *ex contradictione quodlibet* (ECQ). The classical justification for (ECQ) rests on the alleged evidence that no contradiction can be true, which is obviously rejected by dialetheists. In standard natural deduction (ECQ) can be derived using *reductio ad absurdum* (RAA) and other apparently non-problematic rules. (RAA) is immediately rejected by a dialetheist. However, the banish of (RAA) is insufficient to avoid trivialism: Curry's paradox, from which trivialism follows, can be generated without the help of (RAA). In order to save dialetheism from trivialism, in the *Logic of Paradox* (LP) Priest (1979) rejects the general validity of (MPP).

The crucial problem we intend to face in this paper is whether trivialism can follow even from logical principles that are dialetheistically correct. We will first discuss Priest's strategies for avoiding Curry's paradox in LP and in some of its extensions. Then, we will propose a new version of Curry's paradox using the notion of *naïve deducibility*. Such a notion is governed by logical principles that, in our opinion, should be accepted by Priest himself.

2 Curry's paradox and its arithmetical formalization

Curry's paradox belongs to the family of the so-called paradoxes of self-reference (or paradoxes of circularity).[2] In short, the paradox is derived in natural language from sentences like the following:

(b):   If sentence (b) is true, then Santa Claus exists.

Suppose that the antecedent of the conditional in (b) is true, i.e. that sentence (b) is true. Then, by (MPP) Santa Claus exists. So, we have proved the consequent of (b) under the assumption of its antecedent. In other words, we have proved (b). Finally, by (MPP), Santa Claus exists. Of course, we could substitute any arbitrary sentence for 'Santa Claus exists'. As a result, every sentence can be proved and trivialism follows.

In (Priest, 1979, IV, 5) the author observes that, in a semantically closed theory, using (MPP) and absorption, a version of Curry's paradox is derivable. We reconstruct his argument in the language of first order arithmetic with a truth predicate.

Let $\mathcal{L}$ be the language of first order arithmetic and $\mathcal{N}$ its standard model. Let $\mathcal{L}$ be extended to $\mathcal{L}^*$ by introducing a new predicate T.[3] Assume a codification of the syntax of $\mathcal{L}^*$ by natural numbers and extend $\mathcal{N}$ to a model $\mathcal{N}^*$ of $\mathcal{L}^*$ by interpreting T as the truth predicate of $\mathcal{L}^*$, so that, for all $n \in \mathcal{N}$, $T(\underline{n})$ is true *iff* n is the code of a true sentence A of $\mathcal{L}^*$, in symbols $n = \ulcorner A \urcorner$.

Of course, classically, such an interpretation is impossible, because the theory obtained adding to Peano arithmetic the truth predicate for the extended language $\mathcal{L}^*$ with Tarski's biconditionals is inconsistent. Not so for a dialetheist, who loves inconsistent models!

We will show that, if one uses the classical rules of conditional in natural deduction and Tarski's scheme

$$T(\ulcorner A \urcorner) \leftrightarrow A,$$

---

[2]Curry's original paper where the paradox has been introduced is (Curry, 1942).

[3]Unlike Tarski's classical theorem, dialetheism does not exclude the possibility for the truth predicate to be arithmetically expressible. But we do not explore this possibility.

the model $\mathcal{N}^*$ turns out to be trivial. For, let A be any sentence of $\mathcal{L}^*$. By diagonalization, there is a natural number k such that

$$k = \ulcorner T(\underline{k}) \to A \urcorner.$$

We can now derive A as follows (we use $\underline{k}$ as a name of $T(\underline{k}) \to A$):

1	(1)	$T(\underline{k})$	Assumption
1	(2)	$\underline{k}$	1 Tarski's schema
1	(3)	$T(\underline{k}) \to A$	2 Identity
1	(4)	A	1, 3 (MPP)
	(5)	$T(\underline{k}) \to A$	1, 4 $\to$ Introduction
	(6)	$\underline{k}$	5 Identity
	(7)	$T(\underline{k})$	6 Tarski's schema
	(8)	A	5, 7 (MPP)

Priest blocks this derivation in LP by rejecting the general validity of (MPP). According to him, this rule is not valid but *quasi-valid*, i.e. valid insofar as no dialetheia is involved.

We want to discuss this move. We will argue that Priest's rejection of modus ponens (i) is not justified by the presence of dialetheias and (ii) is not in agreement with the intended meaning of the metalinguistic conditional.

3   Criticism to Priest's move

Consider the usual meaning of the conditional in the metalanguage of any mathematical theory. Logicians use the conditional "if A then B" whenever they want to say that the truth of A is a sufficient condition for the truth of B; and that is all. The possible falsity of A or B is not at issue. Sometimes Priest observes that a genuine conditional must preserve falsity from the consequent to the antecedent (see Section 4). Observe, however, that this condition is not implicit in the very meaning of the conditional. It is rather a consequence, in classical logic, of the absence of dialetheias, and hence it should not be endorsed by dialetheism. The metalinguistic meaning of conditional is transferred into the object language by Tarski's clause, according to which "$(A \to B)$ is true" means that if A is true, then B is true. This meaning is in accordance with the deductive practice of mathematicians and is codified in natural deduction by the introduction rule of conditional, according to which one gets a proof of "if A then B" by proving B under the assumption A. Now, if the meaning of the conditional is to be the same both in the language and in the meta-language, as Priest rightly claims, it is incontestable that (MPP) is sound, since it preserves truth quite independently of the possibility that some dialetheia occurs.

Priest tries to escape this objection by identifying $(A \to B)$ with $(\neg A \vee B)$. This identification justifies the rejection of (MPP) by dialetheism. For, if A is a *dialetheia*, $(\neg A \vee B)$ can be true even if B is not. This shows that (MPP) may fail to preserve truth.

Nevertheless, such a move is inappropriate for a dialetheist: the possibility of A being a dialetheia should lead dialetheists to reject the classical equivalence between $(A \to B)$ and $(\neg A \vee B)$. This equivalence holds in classical logic because the truth of $(\neg A \vee B)$ guarantees that truth is preserved from A to B for the very reason that dialetheias are excluded.

Of course, nothing prevents a dialetheist from *defining* $(A \to B)$ as $(\neg A \vee B)$, but in this way she cannot transfer the meaning of "if … then …" from the metalanguage to the object language, against Priest's claim 2. For these reasons we think that the rejection of the general validity of (MPP) is not justified by the mere presence of dialetheias. It is an *ad hoc* move for avoiding trivialism at the cost of a severe limitation of the expressive power of the logical language.

Moreover, the dialetheist reading of $(A \to B)$ as $(\neg A \vee B)$ undermines the intended meaning of the Tarskian scheme, according to which A and $T(\ulcorner A \urcorner)$ have the same truth value. If $T(\ulcorner A \urcorner) \leftrightarrow A$ is understood as

$$\left(\neg T(\ulcorner A \urcorner) \vee A\right) \wedge \left(\neg A \vee T(\ulcorner A \urcorner)\right),$$

when A is a dialetheia, $T(\ulcorner A \urcorner)$ may have any value.

4   Priest's attempt to recover modus ponens

Priest is aware that the material conditional is inadequate to capture the intended meaning of the genuine conditional. In (Priest, 2006b, p. 83), the author recognizes that "any conditional worth its salt should satisfy the modus ponens principle". In (Priest, 2008,

7.4.6), he extends the language of LP by introducing a new conditional $\supset$, satisfying (MPP), defined by the truth function represented in the following truth table:

$\supset$	1	i	0
1	1	0	0
i	1	i	0
0	1	1	1

(where '1' means "true and true only", '0' means "false and false only", and 'i' means "true and false").

This new conditional is characterized by the condition of preserving truth forwards (i.e. from the antecedent to the consequent) and falsity backwards (i.e. from the consequent to the antecedent). This conditional satisfies (MPP), but cuts off the general validity of the introduction rule for the conditional. Consider, for instance, the following derivation:

1	(1)	A	Assumption
2	(2)	B	Assumption
1, 2	(3)	$(A \wedge B)$	1, 2 $\wedge$ Introduction
2	(4)	$(A \supset (A \wedge B))$	$\supset$ Introduction

Now, if A is only true and B is a dialetheia, then $(A \wedge B)$ is a dialetheia and hence $(A \supset (A \wedge B))$ is only false (because falsity fails to be preserved backwards).

However, our derivation of Curry's paradox in Section 2 is still correct, even if the classical $\rightarrow$ is replaced by $\supset$ (and in Tarski's scheme too). For, the deduction from step 1 to step 4 preserves truth and falsity in the appropriate directions, so that, at step 5, $(T(\underline{k}) \supset A)$ turns out to be true.

Alternatively, one can directly verify, by inspection of the truth table of $\supset$, that, if A and $(A \supset B)$ are equivalent (where equivalence is expressed in terms of $\supset$ and $\wedge$), then B is true.

We conclude that the semantics of $\supset$ allows the replication of Curry's paradox.

5   Entailment

In (Priest, 2006b, ch. 6), in order to save (MPP) and avoid Curry's paradox, the author introduces a more sophisticated notion of conditional, called *entailment connective*. Let $\Rightarrow$ be the symbol for this connective. Priest suggests reading $(A \Rightarrow B)$ as "B follows logically from A". As for $\supset$, he imposes to entailment the condition of preserving truth and falsity towards and backwards respectively. But, he observes, due to the fact that logical consequences are such *necessarily*, also preservation of truth and falsity is required to hold *necessarily*. More precisely, Priest defines the truth conditions of entailment within a *possible worlds semantics*, where every interpretation involves an actual world G and an accessibility relation R, as follows:

a) $(A \Rightarrow B)$ is true at world w *iff*, for every world w' accessible to w, if A is true at w', then so is B, and if B is false at w', then so is A;

b) $(A \Rightarrow B)$ is false at w *iff*, at some world w' accessible to w, A is true and B is false.

The actual world G is supposed to be *omniscient*, i.e. to have access to every world, while R is not required to be reflexive, so that a non-actual world may be not accessible to itself (Priest, 2006b, p. 87).

A is said to be a *semantic consequence* of a set of sentences Σ *iff* the following condition is satisfied: for every interpretation, if all $B \in \Sigma$ are true at G, then so is A.

Let us define $A \Leftrightarrow B$ as $((A \Rightarrow B) \wedge (B \Rightarrow A))$. The semantics given for $\Rightarrow$ avoids Curry's paradox. For, B is not a semantic consequence of $(A \Leftrightarrow (A \Rightarrow B))$, as the following counter-model $\mathcal{M}$ shows.

$\mathcal{M}$ has two worlds, G and w; the latter has access only to G (and not to itself). A is only false at G and only true at w; B is only false at G and at w. It turns out that $(A \Rightarrow B)$ is only false at G and only true at w, whence $(A \Leftrightarrow (A \Rightarrow B))$ is only true at G and at w.

The problem arises whether this semantics is adequate to the intended meaning of entailment. Let us see how Priest defends the omniscience of G and the non-reflexivity of R (Priest, 2006b, p. 87):

> Now, how do we know that all the 'possible worlds' in an interpretation are conceivable by people living under those conditions of G? Simply because we are those people (by definition), and we conceive them. It is we who are theorizing, specifying what interpretations are, and we who can spell out any particular [assignment] v_w. If we were

to live under a different set of conditions, however, there would be no guarantee that we would be able to think all of this. Indeed, had we not evolved, we might have been maladapted to our environment, and might not even, therefore, have been able to conceive properly of the conditions under which we actually lived. G is omniscient, but there is no reason, therefore, why any other world should be omniscient or even reflexive.

We certainly agree to the omniscience of G. But the argument for the non-reflexivity of R seems hardly convincing. For the very reason that it is we who theorize and conceive possible worlds, it is we who reason in our actual world about such counterfactual worlds, and not the unknown inhabitants of those worlds. Now the non-reflexivity of w forces us to evaluate an entailment $(A \Rightarrow B)$ in w disregarding the truth values of A and B at w; and that seems to be quite in disagreement with our interest for w. It seems plain that, when considering w, we are interested in what happens in w. Moreover, our logical rules, for the reason that they are *a priori*, are supposed to hold in any world. In particular, if, as Priest rightly stresses, the validity of (MPP) is a *sine qua non* condition for the entailment, it must hold not only in G but in any world. However, in the above counter-model $\mathcal{M}$, (MPP) fails in w, where A and $(A \Rightarrow B)$ are true, while B is false. It may be instructive to turn back to our deduction of Curry's paradox in Section 1:

1	(1)	$T(\underline{k})$	Assumption
1	(2)	$\underline{k}$	1 Tarski's schema
1	(3)	$T(\underline{k}) \to A$	2 Identity
1	(4)	A	1, 3 MPP
	(5)	$T(\underline{k}) \to A$	1, 4 $\to$ Introduction
	(6)	$\underline{k}$	5 Identity
	(7)	$T(\underline{k})$	6 Tarski's schema
	(8)	A	5, 7 (MPP)

The following question arises: at what step the derivation is blocked, when the conditional is interpreted as entailment? Since Priest accepts the validity of Tarski's scheme and (MPP), he should reject step 5. But if A has been derived from $T(\underline{k})$ by means of valid logical rules, it is hard to deny that $T(\underline{k})$ entails A, and hence that $(T(\underline{k}) \Rightarrow A)$

is logically true. What the counter-model $\mathcal{M}$ rejects is rather step 4: Priest's semantics invalidates (MPP) because entailment may fail to preserve truth forwards in some non-actual world, as it happens in $\mathcal{M}$. Of course, Priest might reply that his definition of semantic consequence requires the preservation of truth only in the actual world, and hence that our deduction from step 1 to step 4 is correct because it concerns actual truth, for which (MPP) holds. Nevertheless, he might continue, the fact that truth and falsity in G are preserved from $T(\underline{k})$ to A does not mean that $T(\underline{k})$ entails A, since entailment requires preservation in all accessible worlds; and therefore the incorrect step is number 5. This view, however, does not do justice to abstract hypothetical reasoning. When reasoning under a certain hypothesis H, we do not know, in general, if H is actually true, but we argue *as if* we knew that it were true; and we consider our reasoning correct even if, as a matter of fact, our hypothesis is false. This means that logical reasoning is to be regarded as correct in *any*, possibly counterfactual, situation where our assumption is true. Thus, if (MPP) is acceptable as a *logical* principle, it has to preserve truth and falsity in every possible world. We conclude that Priest's semantics is in disagreement with the apriority of logic and, in particular, inadequate to recover the *logical* validity of (MPP).

6 A new argument for trivialism

We will propose a new version of Curry's paradox, without making use of (MPP). We use a notion of naïve deducibility that, in our opinion, should be perfectly acceptable by Priest.

In his discussion about Gödel's theorem, Priest defends the notion of *naïve proof* (Priest, 2006b, p. 40):

> Proof, as understood by mathematicians (not logicians), is that process of deductive argumentation by which we establish certain mathematical claims to be true. In other words, suppose we have a mathematical assertion, say a claim of number theory, whose truth or falsity we wish to establish. We look for a proof or a refutation, that is a proof of its negation... I will call the informal deductive arguments from basic statements *naïve proofs*.

Priest holds that naïve proof is inconsistent but correct, since it is the official means for recognizing mathematical truths. According to him, inconsistency arises from the presence of dialetheias. Furthermore, he argues for the thesis that the notion of naïve proof of an arithmetical sentence is expressible in the arithmetical language. We do not endorse the latter highly problematic thesis. Instead, consider the extension $\mathcal{L}^*$ of the language $\mathcal{L}$ of first order arithmetic, obtained by introducing the predicate $P(x)$. Extend the standard model $\mathcal{N}$ of arithmetic to the model $\mathcal{N}^*$, where P is interpreted as naïve provability for the language $\mathcal{L}^*$ (with reference to a numerical codification of the syntax of $\mathcal{L}^*$).[4] More precisely, $P(\ulcorner A \urcorner)$ means: it is naïvely provable that A is true in $\mathcal{N}^*$. According to Priest, $P(x)$ satisfies the following principles:

1. $P(\ulcorner A \urcorner) \rightarrow A$ is naïvely provable,

2. If A is naïvely provable, then $P(\ulcorner A \urcorner)$ is naïvely provable.

Priest justifies these principles by observing that (Priest, 2006b, p. 238):

> for (1) it is analytic that whatever is naïvely provable is true. Naïve proof is just that sort of mathematical argument that establishes something as true. And since this is analytic, it is itself naïvely provable... For (2), if something is naïvely proved then this fact itself constitutes a proof that A is provable.

Similarly, we extend $\mathcal{L}$ to a language $\mathcal{L}^*$ by introducing a binary predicate of naïve deducibility $D(x, y)$. Then we extend the standard model $\mathcal{N}$ of arithmetic to the model $\mathcal{N}^*$ of $\mathcal{L}^*$, where D is interpreted as the naïve deducibility relation for $\mathcal{L}^*$, that is, $D(x, y)$ means "y is naïvely deducible from x". More explicitly, $D(x, y)$ means that *there is a naïve proof that, assuming that x is true in $\mathcal{N}^*$, leads to the conclusion that y is true in $\mathcal{N}^*$*.

In a system of natural deduction, the analogous principles to (1) and (2) above are the following elimination and introduction rules for D:

---

[4] If, as Priest maintains, P is arithmetically expressible, then $\mathcal{L} = \mathcal{L}^*$.

- (DE): From premises A and $D(\ulcorner A \urcorner, \ulcorner B \urcorner)$ you can derive B. The conclusion depends on all assumptions the premises depend on.

- (DI): From premise B, depending on the unique assumption A, one can infer $D(\ulcorner A \urcorner, \ulcorner B \urcorner)$, discharging A.

It is worthwhile comparing (DE) with *modus ponens*. Suppose that A is a *dialetheia*. While the conditional $(A \to B)$, when identified as $(\neg A \lor B)$, may fail to preserve the truth from the antecedent to the consequent because the inference from A and $(\neg A \lor B)$ to B is blocked, $D(\ulcorner A \urcorner, \ulcorner B \urcorner)$ preserves the truth from A to B, because the inference from A and $D(\ulcorner A \urcorner, \ulcorner B \urcorner)$ to B is not blocked. Indeed, according to the intended meaning of D, if $D(\ulcorner A \urcorner, \ulcorner B \urcorner)$ is true, the truth of B is provable under the mere assumption that A is true (the fact that A also may be false is not at issue here).

Regarding (DI), if B is derived from the unique assumption A, since the soundness of inference rules is recognized by informal reasoning, there is a naïve reasoning that leads from A to B.

Now, Let A be any $\mathcal{L}^*$-sentence. By diagonalization, we get a natural number k such that

$$k = \ulcorner D(\underline{k}, \ulcorner A \urcorner) \urcorner.$$

We can prove A as follows:

Proof. Using $\underline{k}$ as a name of $D(\underline{k}, \ulcorner A \urcorner)$, suppose that $\underline{k}$ is true. Since $\underline{k}$ says that A is deducible from $\underline{k}$ and deduction is sound, A is true. So we have proved A from the assumption $\underline{k}$. Hence $D(\underline{k}, \ulcorner A \urcorner)$, i.e. $\underline{k}$, is true. And, since deduction is sound, A is true. $\qquad\square$

A *formal proof* of A (where $\underline{k}$ is used again a name of $D(\underline{k}, \ulcorner A \urcorner)$):

1	(1)	$\underline{k}$	Assumption
1	(2)	$D(\underline{k}, \ulcorner A \urcorner)$	1 Identity
1	(3)	A	1, 2 (DE)
	(4)	$D(\underline{k}, \ulcorner A \urcorner)$	1, 3 (DI) (discharging (1))
	(5)	$\underline{k}$	4 Identity
	(6)	A	4, 5 (DE)

Since A is arbitrary, $\mathcal{N}^*$ is trivial. But $\mathcal{N}^*$ differs from $\mathcal{N}$ only for the introduction of the relation of naïve deducibility: the arithmetical sentences of $\mathcal{L}$ are interpreted in $\mathcal{N}^*$ as in $\mathcal{N}$. Therefore $\mathcal{N}$ is trivial as well.

7 Conclusion

We conclude that, according to his perspective, Priest should recognize that the standard model of arithmetic is trivial. After all, why should trivialism not be accepted as a *fact of life*? However, never mind! The triviality of arithmetic does not invalidate the proof by induction that every arithmetical sentence has a unique truth-value. Thus dialetheists should not be afraid of trivialism. Nevertheless, their philosophical assumptions allow them to hold that arithmetic is trivial, even if it is not.

References

Berto, F. (2007). *How to Sell a Contradiction: The Logic and Metaphysics of Inconsistency*. London: College Publications.

Curry, H. B. (1942). The Inconsistency of Certain Formal Logics. *Journal of Symbolic Logic, 7*, 115–117.

Fine, K. (1975). Vagueness, truth and logic. *Synthese, 30*, 265–300.

Priest, G. (1979). The Logic of Paradox. *Journal of Philosophical Logic, 8*, 219–241.

Priest, G. (2002a). *Beyond the limits of though*. Oxford: Oxford University Press.

Priest, G. (2002b). Paraconsistent Logic. In D. Gabbay & F. Guenthner (Eds.), *Handbook of philosophical logic* (second ed., Vol. 6, p. 287-393). Dordrecht: Kluwer Academic Publisher.

Priest, G. (2006a). *Doubt Truth to be a Liar*. Oxford: Oxford University Press.

Priest, G. (2006b). *In Contradiction*. Oxford: Oxford University Press.

Priest, G. (2008). *An Introduction to Non-Classical Logic*. Cambridge: Cambridge University Press.

Massimiliano Carrara, Silvia Gaio, and Enrico Martino
Department of Philosophy, University of Padua—Italy
P.zza Capitaniato 3, I-35139 Padova
e-mail: {massimiliano.carrara,silvia.gaio,enrico.martino}@
unipd.it

A Prologue to the Theory of Deduction

Kosta Došen[*]

Abstract

This is an introduction to general proof theory in terms of some philosophical considerations. The central problem of this theory is the problem of identity criteria for deductions. When it is treated in categorial proof theory, this problem is closely tied to categorial coherence results, which may be understood as completeness results for systems of equations between deductions with respect to simple model categories. With the help of the categorial notion of isomorphism, the categorial perspective enables us to formulate in precise mathematical terms criteria for propositional identity.

1 Introduction

The central question of general proof theory as defined by Prawitz in (Prawitz, 1971) may be understood as the question "What is a deduction?", and the central problem of this theory is the problem of identity criteria for deductions. This problem is treated either in the typed lambda calculus or in terms of category theory. These two approaches are up to a point convergent, the latter one providing sometimes a deeper perspective. The categorial approach to proof theory—categorial proof theory—is also closely tied to what in category theory is called coherence results, which may be understood as completeness results for systems of equations between deductions with

[*]*Acknowledgements.* The author is grateful to Vladimír Svoboda for inviting him to deliver a talk based on this paper at *Logica 2010* (after his coming to *Logica 2009* had to be cancelled). He is also grateful to Zoran Petrić for reading the first draft, and to Senka Milošević for drawing his attention to the objectivity of deduction mentioned in Section 3. Work on the paper was supported by the Ministry of Science of Serbia (Grant ON174026).

respect to manageable models, often of a graphical kind. With the
help of the categorial notion of isomorphism, the categorial perspective enables us to formulate in precise mathematical terms criteria
for propositional identity, i.e. identity of meaning for propositions.
This paper consists in some introductory philosophical considerations
concerning these matters of general proof theory.

2   Proofs and deductions

The word "proof" may refer to a proof with no hypothesis (or assumption). This is what this word does in its marked usage. In its
unmarked usage, it may refer also to a hypothetical proof, i.e. a proof
from hypotheses. (In the terminology of the Prague Linguistic Circle,
"fox" in its *marked* usage refers to a male animal, the male partner
of a vixen, while in its *unmarked* usage it refers to a male or female
animal.)

General proof theory ought to be concerned with proofs in the
unmarked usage. In order to make this clear, it is wiser to opt for the
synonymous term "deduction", which cannot be mistaken for "proof"
in the marked usage. So general proof theory ought to be about
deductions.

That this precaution is not exaggerated is corroborated by the fact
that proof-theoretical investigations of intuitionistic logic are inclined
to understand "proof" in the marked usage. This is shown by the
Curry-Howard correspondence.

There we find typed lambda terms t as codes of natural deduction
derivations. If t codes the derivation that ends with the formula B as
the last conclusion, then this may be written $t : B$, and we say that
t is of *type B*. (Formulae are of course of the grammatical category
of propositions.) Our derivation may have uncancelled hypotheses.
That will be seen by t's having possibly a free variable x, which codes
an occurrence of a formula A as hypothesis; i.e. we have $x : A$, an x of
type A.

All this makes conclusions prominent, while hypotheses are veiled.
Conclusions are clearly there to be seen as types of terms, while hypotheses are hidden as types of free variables, which are cumbersome
to write always explicitly when the variables occur as proper subterms
of terms. The desirable terms are closed terms, which code derivations

where all the hypotheses have been cancelled. These closed terms are supposed to play a key role in understanding intuitionistic logic.

Another asymmetry is brought by the usual format of natural deduction, where there can be more than one premise, but there is always a single conclusion. This format favours intuitionistic logic, and in this logic the coding with typed lambda terms works really well with implication and conjunction, while with disjunction there are problems.

An alternative to this coding would be a coding of derivations that would allow hypotheses to be as visible as conclusions, and to be treated on an equal footing also with respect to multiplicity. With such a coding we could hope to deal too with classical logic, with all its Boolean symmetries, and with disjunction as well as with conjunction.

Such an alternative coding exists in categorial proof theory. There one writes $f\colon A \vdash B$ as a code for a derivation that starts with premise A and ends with conclusion B. The arrow term f is an arrow term of a category—a *cartesian closed category* if we want to cover the conjunction-implication fragment of intuitionistic logic. The *type* of f is now not a single formula, but an ordered pair of formulae (A, B). The notation $A \vdash B$ serves just to have something more suggestive than (A, B). (In categories one usually writes $f\colon A \to B$ instead of $f\colon A \vdash B$, but $\to$ is sometimes used in logic for implication, and we should not be inclined to confuse deduction with implication just because of notation; cf. Section 4 below.)

If B happens to be derived with all hypotheses cancelled, then we will have $f\colon \top \vdash B$, with the constant formula $\top$ standing for the empty collection of hypotheses. If we happen to have more than one hypothesis, but as usual a finite number of them, then we will assume that with the help of conjunction all these hypotheses have been joined into a single hypothesis. So the categorial notation $f\colon A \vdash B$ with a single premise does not introduce a cramping limitation; at least not for the things intended to be said here.

The typed lambda coding of the Curry-Howard correspondence, involving finite product types and functional types, and the categorial coding in cartesian closed categories are equivalent in a very precise sense. This has been first shown by Lambek (see Lambek, 1974; Lambek & Scott, 1986; Došen, 1996, 2001). The import of the two formalisms is however not exactly the same. The typed lambda calculus may suggest something different about the subject matter than category theory. It makes prominent the *proofs* $t: B$—and we think immediately of the marked ones, without hypotheses—while category theory is about the *deductions* $f: A \vdash B$.

Logic is concerned not with any deductions, but first of all with *formal* deductions. Perhaps it is concerned only with such deductions. These are deductions that hold in virtue of the *form* exhibited by keeping constant some particular expressions—the logical constants (the standard list is made of the connectives and quantifiers of first-order logic plus the binary predicate of equality)—and replacing everything else by variables, of appropriate grammatical categories. The expressions of the grammatical category of propositions that exhibit this form are called *formulae*.

If general proof theory is described as the theory of formal deduction, it might seem it would cover the whole field of logic, as traditionally conceived. Formal deductions are however hardly the subject matter of model theory, computability theory and set theory. (In Hodges, 2007 it is found that model theory could fall under the heading of *theory of definition*, which should be a traditional concern of logicians.) The theory of formal deduction would not cover the whole field of logic as it is conceived since the last century. Once it is properly developed, it could perhaps pretend to occupy a place more central than today.

3   Deductions in general

Following Prawitz's suggestion in (Prawitz, 1971), the primary concern of general proof theory should then be the question "What is a deduction?". Let us consider this question in all its generality. So we are not concerned only with formal deductions, where premises and conclusions are formulae, but with any deductions, where premises and conclusions are propositions. What is then a deduction in general, in this broad sense?

There is a strong tendency to answer this question by relying on the notion of proposition as more fundamental. It is as if Frege's recommendation from the *Grundlagen der Arithmetik* to look after meaning in the context of a proposition was understood to apply not only to bits of language narrower than propositions, which should be placed in the broader propositional context, but also to something broader than propositions, as deductions, in which propositions partake, which should be explained in terms of the narrower notion of proposition.

A deduction is usually taken as something involving propositions. Restricting ourselves to deductions with single premises, as we agreed to do above, for ease of exposition, we may venture to say that a deduction consists in *passing* from a proposition called premise to a proposition called conclusion.

What could "passing" mean here? Another teaching of Frege from *Grundlagen der Arithmetik* would not let us understand this passing as something happening in our head. Such an understanding would expose us to being accused of psychologism. No, this passing should be something objective, something done or happening independently of any particular thinking subject, something sanctioned by language and the meaning it has.

The temptation of psychologism is particularly strong here, but as a proposition is not something mental that comes into being when one asserts a sentence, so a deduction should not be taken as a mental activity of passing from sentences to sentences or from propositions to propositions. Such a mental activity exists, as well as the accompanying verbal and graphical activities, but the deduction we are interested in is none of these activities. It is rather something in virtue of which these activities are judged to be correctly performed or not. It is something tied to rules governing the use of language, something based on these rules, which are derived from the meaning of language, or which confer meaning to it.

A deduction could be compared to a move in a game like chess. We envisage here possible moves in particular situation on the board in a particular game. Such moves are not moves in the physical world with the help of brain and fingers. Moves in the physical world may also be incorrectly performed. We are however concerned with the correct abstract moves, sanctioned by the rules of the game, whose existence is objective, and which do not owe anything to particular chess players. There are such moves that no chess player has ever envisaged.

This chess terminology suggests saying, as a Wittgenstein would perhaps do, that deduction is a *move* in a language game. However, "move" does not fare much better than "passing". One smacks of psychologism, but the other is very metaphorical, and neither is very explanatory.

4   Deductions and consequence relations

Can deductions be reduced to consequence relations? So that having a deduction from A to B means just that B is a consequence of A. This would square well with the objective character of deductions we have just talked about, because B's being a consequence of A is something objective. "Consequence" should presumably be understood here in a syntactical manner, but because of the completeness of classical first-order logic it could even mean semantical consequence. If our view of logic is dictated by semantics, as it is understood in model-theoretical terms, then the objectivity of consequence has semantical grounds.

Since B is a consequence of A whenever the implication $A \to B$ is true or correct, there would be no essential difference between the theory of deduction and the theory of implication. A deduction is often written vertically, with the premise above the conclusion,

$$\frac{A}{B}$$

and an implication is written horizontally $A \to B$, but besides that, and purely grammatical matters, there would not be much difference.

This reduction of deduction to implication accords well with the point of view, which we mentioned above, where propositions are taken as a more fundamental notion. And this is indeed the point of view of practically all of the philosophy of logic and language in the twentieth century. Propositions do not only play the leading role in language, but everything is reduced to them.

This applies not only to classically minded theories where the essential, and desirable, quality of propositions is taken to be truth, but also to other theories, like constructivism in mathematics, or verificationism in science, where this quality is something different. It may be deemed strange that even in constructivism, where the quality is often described as *provability*, deductions are not more prominent. Rather than speak about deductions, constructivists, such as intuitionists, tend to speak about something more general covered by the magical word "construction". (Constructions produce mathematical objects as well as proofs of mathematical propositions, which are about these objects.) Where above we spoke about *passing* and *moves*, a constructivist would presumably speak about *constructions*.

By reducing deductions from A to B to ordered pairs (A, B) in a consequence relation we would loose the need for the categorial point of view. The f in $f: A \vdash B$ would become superfluous. There would be at most one arrow with a given source and target, which means that our categories would be preordering relations (i.e. reflexive and transitive relations). These preorderings are consequence relations.

With that we would achieve something akin to what has been achieved for the notion of function. This notion has been *extensionalized*. It has been reduced to a set of ordered pairs. If before one imagined functions as something like a *passing* from an argument to the value, or a *move* from an argument to the value, now a function is just a set of ordered pairs made of arguments and values. Analogously, deductions would be the ordered pairs made of premises and conclusions.

This extensionalizing might represent a progress, but it has left in its vicinity, in the theory of computing, an unresolved matter concerning the notion of algorithm. While it might be desirable to extensionalize the notion of function, is it desirable in the same manner to extensionalize the notion of algorithm? Are all algorithms with the same input and output the same? Common sense says no, but no developed or widely accepted theory exists concerning this matter (see Moschovakis, 2001 and Blass, Dershowitz, & Gurevich, 2009).

Gentzen's results in general proof theory concerning the normal forms of natural deduction and sequent derivations demonstrate the technical advantage of taking deductions seriously. Not so much as something irreducible to a consequence relation, but as something vertical, certainly irreducible to the horizontal implication.

5   Identity of deductions

The extensionalizing of the notion of deduction which consists in its reduction to the notion of consequence relation can be called into question if we are able to produce examples of two different deductions with the same premise and the same conclusion. Here are first two such examples of formal, logical, deductions, which involve conjunction, the simplest and most basic of all logical connectives.

From $p \wedge p$ to p there are two deductions, one obtained by applying the first rule of conjunction elimination, the first projection rule,

$$\frac{A \wedge B}{A}$$

and the other obtained by applying the second projection rule

$$\frac{A \wedge B}{B}$$

The other example is given by the two deductions from $p \wedge p$ to $p \wedge p$, one being the identity deduction, an instance of

$$\frac{A \wedge B}{A \wedge B}$$

and the other an instance of permutation

$$\frac{A \wedge B}{B \wedge A}$$

These and other such examples from logic redeem the categorial point of view (see Došen & Petrić, 2004, 2007). In the first example we have $\pi^1 : p \wedge p \vdash p$ and $\pi^2 : p \wedge p \vdash p$ with $\pi^1 \neq \pi^2$, and in the second $\mathbf{1} : p \wedge p \vdash p \wedge p$ and $c : p \wedge p \vdash p \wedge p$ with $\mathbf{1} \neq c$.

A category where these arrows are exemplified is $\mathcal{C}$, which is the category with binary product freely generated by a set of objects. The category $\mathcal{C}$ models deductions involving only conjunction. It does so for both classical and intuitionistic conjunction, because the deductions involving this connective do not differ in the two alternative logics. This is a common ground of these two logics.

The notion of binary product codified in $\mathcal{C}$ is one of the biggest successes of category theory. The explanation of the extremely important notion of ordered pair in terms of this notion is the most convincing corroboration of the point of view that mathematical objects should be characterized only up to isomorphism. It is remarkable that the same matter should appear at the very beginning of what category theory has to say about deductions in logic, in connection with the connective of conjunction.

For the category $\mathcal{C}$ there exists a kind of completeness theorem, which categorists call a *coherence* result. There is namely a faithful functor from $\mathcal{C}$ to the model category $\mathcal{M}$, which is the opposite of the category of finite ordinals with functions as arrows. With this functor, π^1 and π^2 above correspond respectively to

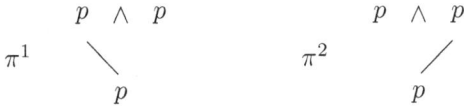

$$\pi^1 \quad \begin{array}{c} p \;\wedge\; p \\ \searrow \\ p \end{array} \qquad\qquad \pi^2 \quad \begin{array}{c} p \;\wedge\; p \\ \nearrow \\ p \end{array}$$

while $\mathbf{1}$ and c correspond respectively to

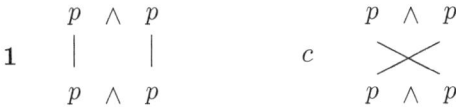

$$\mathbf{1} \quad \begin{array}{c} p \;\wedge\; p \\ | \quad | \\ p \;\wedge\; p \end{array} \qquad\qquad c \quad \begin{array}{c} p \;\wedge\; p \\ \times \\ p \;\wedge\; p \end{array}$$

Another example of two different formal deductions with the same premise and the same conclusion, which involves graphs of a slightly more complicated kind, is given by

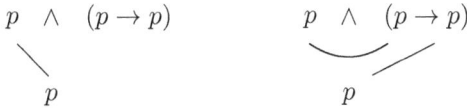

$$\begin{array}{c} p \;\wedge\; (p \to p) \\ \searrow \\ p \end{array} \qquad\qquad \begin{array}{c} p \;\wedge\; (p \to p) \\ \smile \quad \diagup \\ p \end{array}$$

The first deduction is made by conjunction elimination, while the second by modus ponens.

Finally we give an example where the deductions are not formal in the logical sense (at least they are not such at first glance, before further analyzing). Whoever has heard of Achilles, and knows what "faster" means, knows that (1) Achilles is faster than anybody else, and whoever has heard of the Tortoise, and knows what "slower" means, knows that (2) the Tortoise is slower than anybody else. Then from the premise that Achilles and the Tortoise have started their famous race one may deduce (*pace* Zeno) that Achilles will overtake the Tortoise. Two different deductions however lead from our premise to this conclusion: one relies on (1), and the other on (2).

Coherence is one of the main inspirations of categorial proof theory (see Došen & Petrić, 2004). The other, older, inspiration, which works for deductions in intuitionistic logic, comes ultimately from the notion of adjunction (see Lawvere, 1969; Došen, 1999, 2001, 2006).

In model categories such as we find in coherence results we have models of equational theories axiomatizing identity of deductions. These are not models of the theorems of logic. This is often unclear, partly because of uncertainties related to matters considered in Section 2. Moreover, some authors seem to think that models are better if they are complicated. (They may indeed be more likely to impress the public, and hence be better for marketing purposes.) The primary purpose of these models is to serve to obtain simple decision procedures for identity of deductions. So they should be manageable and as simple as possible, as the model category $\mathcal{M}$ above is.

The arrows of the model categories are hardly what deductions really are. It is not at all clear that these categories provide a real semantics of deductions (cf. Došen, 2006).

Invoking now another principle of Frege's *Grundlagen der Arithmetik*, we might look for an answer to the question "What is a deduction?" by looking for a criterion of identity between deductions. We would strive to define a significant and plausible equivalence relation on derivations as coded by arrow terms of our syntactical categories, and equivalence classes of derivations, or equivalence classes of arrow terms, which are the arrows of our syntactical categories, would stand for deductions.

A deduction $f \colon A \vdash B$ would be something sui generis, that does not reduce to its type, the ordered pair (A, B). It would be represented by an arrow in a category, to which is tied a criterion of identity given by the system of equations that hold in the category. The category should not be a preorder.

In the arrow term f of $f \colon A \vdash B$, the rules of inference, the rules of deduction, involved in building a deduction are made manifest as operations for building this term. The theory of deduction is as a matter of fact the theory of such operations (usually partial). It is an algebraic theory codifying with equations the properties of these operations.

It is rather to be expected that the theory of deduction should be the theory of the rules of deduction, as arithmetic, the theory of natural numbers, is the theory of arithmetic operations (addition, multiplication etc.). Extensionalizing the notion of deduction by reducing it to consequence (as in the preceding section) makes us forget the rules of deduction, which are prominent in categorial proof theory.

This would be the way to approach the question "What is a deduction?", and it is a mathematical way. Let us consider some philosophical aspects of this mathematical approach.

From a classical point of view, the desirable quality of propositions, their correctness, is their truth. If the notion of deduction is something sui generis, not reducible to the notion of proposition, the desirable quality of deductions, their correctness, need not be reducible to the desirable quality of propositions, which from a classical point of view is their truth, as we have just said. A correct deduction would not be just one that preserves truth—a correct consequence relation is that. A deduction $f: A \vdash B$ is not just $A \vdash B$; we also have the f. As a matter of fact, the deduction is f. It is a necessary condition for a correct $f: A \vdash B$ that B be a consequence of A, but this is not sufficient for the correctness of f. This is not what the correctness of f consists in. The correctness of a deduction would be as the notion of deduction itself something sui generis, not to be explained by the desirable quality of propositions.

We might perhaps even try to turn over the positions, and consider that the desirable quality of propositions should be explained in terms of the correctness of deductions. This is presumably congenial to a point of view like that found in intuitionism, where the desirable quality of propositions is taken to be provability, i.e. deducibility from an empty collection of premises. We could however take an analogous position with a classical point of view, where the *truth* of analytic propositions would be guaranteed by the correctness of some deductions (cf. Dummett, 1991, p. 26). The correctness of the deductions underlies the truth, and not the other way round.

If deductions are objective, as we have recommended in Section 3, do incorrect deductions exist? The subjective activity of deducing may be performed incorrectly, but how can the objective paradigm be incorrect? This question seems to be analogous to the question whether propositions that are not true exist objectively. Does the proposition $2 + 2 = 5$ exist objectively? These questions have a typically philosophical, suspect, ring; with them we might be stepping into an ontological quagmire. Without trying to survey this whole, dangerous, ground, let us try to see what difference there may be here between deductions and propositions.

A correct deduction is like a well-formed syntactical object—for example, a formula—and an incorrect one is like a badly formed syntactical object. (In logic, derivations, like other syntactical objects, are defined inductively, and deductions are equivalence classes of derivations.) An incorrect deduction does not exist as a badly formed syntactical object does not belong to the language. On the other hand, $2 + 2 = 5$ is not badly formed syntactically; it is a well-formed formula. Such a formula may be considered incorrect for syntactical reasons, because it is not a theorem, i.e. because it is underivable in a formal system. From a classical point of view, it is however usually branded as incorrect, as being untrue, for semantical reasons. For semantical reasons, other formulae would be considered correct, i.e. true.

It seems that a deduction $f \colon A \vdash B$ too may be considered incorrect for semantical reasons. This may be done simply because B is not a consequence of A. It is not clear whether this may be done also because f does not satisfy something required by a putative semantics of deduction, something formulated perhaps in terms of models of deduction, like our category $\mathcal{M}$ above.

Can f be branded as correct for semantical reasons? We said already that this may not be done simply because B is a consequence of A. Are there then other semantical reasons that might induce us to do that? This question is unclear. Many things are unclear about the semantics of deduction (see Došen, 2006). Many things are however unclear about semantics in general.

6   Isomorphism of propositions

We will now consider another philosophical aspect of our mathematical approach to the question "What is a deduction?". This is something that may be interesting for semantics. It has to do with the meaning of propositions.

Isomorphism between formulae should be an equivalence relation stronger than mutual implication. This is presumably the relation underlying the relation that holds between propositions that have the same meaning just because of their logical form. Any propositions that are instances, with the same substitution, of isomorphic formulae would have the same meaning, which presumably need not be the case for formulae that are just equivalent, i.e. which just imply each other.

One may try to characterize isomorphic formulae by looking only into the inner structure of formulae . This is the way envisaged by Carnap and Church (see Carnap, 1947, Sections 14–15, where related work by Quine and C.I. Lewis is mentioned, Church, 1950; Anderson, 1998, Section 2, and references therein).

Another way is to try to characterize isomorphism between formulae by looking also at the outer structure in which formulae are to be found. This outer structure may be a deductive structure, characterized in terms of categories in categorial proof theory. These are the syntactical categories envisaged in the preceding sections of this paper: their objects are formulae and their arrows are deductions.

Isomorphism between formulae may then be understood exactly as isomorphism between objects is understood in category theory. The formulae A and B are isomorphic when there is a deduction $f : A \vdash B$, and another deduction $g : B \vdash A$, such that f composed with g, i.e. $g \circ f : A \vdash A$ is equal to the identity deduction from A to A, namely $1_A : A \vdash A$, while $f \circ g : B \vdash B$ is equal to $1_B : B \vdash B$. This analysis of isomorphism presupposes a criterion of identity for deductions, which is formalized by equality between arrows in our syntactical categories.

That A and B are isomorphic means here intuitively that they function in the same manner in deductions. In a deduction one can replace one by the other, either as premise or as conclusion, so that nothing is lost, nor gained. The replacements, which are made by composing our deduction with the deductions f and g, are such that they enable us to return to our original deduction by further composing with g and f, since $g \circ f$ and $f \circ g$ are identity deductions, and hence may be cancelled. (For this view concerning isomorphic formulae and its relationship with propositional identity, see Došen, 1997, Section 9; Došen, 2006, Section 5; Došen & Petrić, 2009.)

The study of isomorphic formulae first started in intuitionistic logic, for which it is widely believed that we have solid nontrivial criteria of identity of deductions. These criteria are provided either in terms of equality in the typed lambda calculus, via the Curry-Howard correspondence, or in terms of equality between arrows in categories based on cartesian closed categories (as we mentioned in Section 2).

A result exists in this area for the conjunction-implication fragment of intuitionistic logic (see Soloviev, 1981), but the problem of characterizing formulae isomorphic in the whole of intuitionistic propositional logic (which is related to Tarski's high-school algebra problem;

see Burris & Lee, 1993) seems to be still open. There is a further result characterizing isomorphic formulae in the analogous multiplicative fragment of linear logic, which corresponds to symmetric monoidal closed categories, and is common to classical and intuitionistic linear logic (see Došen & Petrić, 1997).

Results characterizing isomorphic formulae in classical propositional logic may be found in (Došen & Petrić, 2009). These results cover also a fragment of classical linear propositional logic. For these results to be significant, we need for the logics we want to cover a plausible and nontrivial notion of equality of arrows in categories formalizing identity of deductions in these logics. A consensus for classical linear propositional logic may be found around the multiplicative fragment of that logic caught by proof nets, which leads to notions of category closely related to star-autonomous category (see Došen & Petrić, 2007 and references therein).

For classical propositional logic, it is on the contrary widely believed that no nontrivial notion of category would do the job. It is believed that no nontrivial notion of Boolean category may be found. This is indeed the case if one wants these Boolean categories to be cartesian closed (see Došen, 2003, Section 5; Došen & Petrić, 2004, Section 14.3, and references therein). But, whereas on the level of theorems classical logic is an extension of intuitionistic logic, it is not clear that this should be so at the level of deductions and of criteria concerning their identity.

If one does not require that Boolean categories be cartesian closed categories, and bases identity of deductions in Boolean categories on coherence results analogous to those available for classical linear propositional logic, a nontrivial notion may arise. The coherence results in question are categorial results analogous to the classical coherence result of Kelly and Mac Lane for symmetric monoidal closed categories (see Kelly & Mac Lane, 1971). They reduce equality of arrows in the syntactical category to equality of arrows in a graphical model category.

Such a nontrivial notion of Boolean category may be found in (Došen & Petrić, 2004, Chapter 14), and (Došen & Petrić, 2009) deals with isomorphism of formulae engendered by that particular notion, and by another notion motivated more consistently by generality of deductions (see Došen, 2003). These results give a complete characterization of isomorphic formulae in classical and classical linear propositional logic. These characterizations are such that they easily lead to decision procedures for the isomorphisms in question.

References

Anderson, C. (1998). Alonzo Church's contributions to philosophy and intensional logic. *The Bulletin of Symbolic Logic*, *4*, 129–171.

Blass, A., Dershowitz, N., & Gurevich, Y. (2009). When are two algorithms the same? *The Bulletin of Symbolic Logic*, *15*, 145–168.

Burris, S., & Lee, S. (1993). Tarski's high school identities. *American Mathematical Monthly*, *100*, 231–236.

Carnap, R. (1947). *Meaning and Necessity: A Study in Semantics and Modal Logic*. Chicago: The University of Chicago Press.

Church, A. (1950). On Carnap's analysis of statements of assertion and belief. *Analysis*, *10*, 97–99.

Došen, K. (1996). Deductive completeness. *The Bulletin of Symbolic Logic*, *2*, 243–283, 523. (for corrections see Došen, 1999, Section 5.1.7 and Došen, 2001)

Došen, K. (1997). Logical consequence: A turn in style. In M. L. Dalla Chiara, K. Doets, D. Mundici, & J. van Benthem (Eds.), *Logic and scientific methods* (Vol. 1, pp. 289–311). Dordrecht: Kluwer.

Došen, K. (1999). *Cut elimination in categories*. Dordrecht: Kluwer.

Došen, K. (2001). Abstraction and application in adjunction. In Z. Kadelburg (Ed.), *Proceedings of the Tenth Congress of Yugoslav Mathematicians* (pp. 33–46). Belgrade: Faculty of Mathematics, University of Belgrade. (available at: `http://arXiv.org`)

Došen, K. (2003). Identity of proofs based on normalization and generality. *The Bulletin of Symbolic Logic*, *9*, 477–503. (version with corrected remark on difunctionality available at: `http://arXiv.org`)

Došen, K. (2006). Models of deduction. *Synthese*, *148*, 639–657.

Došen, K., & Petrić, Z. (1997). Isomorphic objects in symmetric monoidal closed categories. *Mathematical Structures in Computer Science*, *7*, 639–662.

Došen, K., & Petrić, Z. (2004). *Proof-Theoretical Coherence.* London: KCL Publications (College Publications). (revised version of 2007 available at: http://www.mi.sanu.ac.rs/~kosta/coh.pdf)

Došen, K., & Petrić, Z. (2007). *Proof-Net Categories.* Monza: Polimetrica. (preprint of 2005 available at: http://www.mi.sanu.ac.rs/~kosta/pn.pdf)

Došen, K., & Petrić, Z. (2009). *Isomorphic formulae in classical propositional logic.* (preprint available at: http://arXiv.org)

Dummett, M. (1991). *Frege: Philosophy of Mathematics.* London: Duckworth.

Hodges, W. (2007). The scope and limits of logic. In D. Jacquette (Ed.), *Philosophy of Logic* (pp. 41–65). Amsterdam: North-Holland.

Kelly, G., & Mac Lane, S. (1971). Coherence in closed categories. *Journal of Pure and Applied Algebra, 1*, 97–140, 219.

Lambek, J. (1974). Functional completeness of cartesian categories. *Annals of Mathematical Logic, 6*, 259–292.

Lambek, J., & Scott, P. (1986). *Introduction to Higher-Order Categorical Logic.* Cambridge: Cambridge University Press.

Lawvere, F. (1969). Adjointness in foundations. *Dialectica, 23*, 281–296.

Moschovakis, Y. (2001). What is an algorithm? In B. Engquist & W. Schmid (Eds.), *Mathematics unlimited—2001 and beyond* (pp. 919–936). Berlin: Springer.

Prawitz, D. (1971). Ideas and results in proof theory. In J. Fenstad (Ed.), *Proceedings of the second scandinavian logic symposium* (pp. 235–307). Amsterdam: North-Holland.

Soloviev, S. (1981). The category of finite sets and cartesian closed categories (in Russian). *Zapiski Nauchnykh Seminarov (LOMI), 105*, 174–194. (English translation in *Journal of Soviet Mathematics, 22*, 1983, 1387–1400)

Kosta Došen
Faculty of Philosophy, University of Belgrade, and
Mathematical Institute, SANU
Knez Mihailova 36, p.f. 367, 11001 Belgrade, Serbia
e-mail: kosta@mi.sanu.ac.rs

Transparent Quantification into Hyperintensional Contexts

Marie Duží* and Bjørn Jespersen*

Abstract

We describe an extensional logic of hyperintensions (Tichý's Transparent Intensional Logic: TIL) that comes with a context-invariant semantics, a ramified type hierarchy, partial functions, and hyperintensions that may fail to yield an object. TIL preserves transparency and compositionality for all contexts, does not turn attitude reports into *oratio obliqua*, and so validates quantifying into hyperintensional contexts. Section 2 summarizes the relevant semantic foundations. Section 3 sets out the relevant logical foundations. Section 4 proves a general rule of quantifying into hyperintensional contexts.

1 Introduction

The rule of existential generalization is a defining feature of extensional logic. The conclusion simply makes explicit an existential commitment incurred by the premise:

$$\frac{Fa}{\exists x(Fx)}$$

*Versions of this paper were read by Marie Duží at LOGICA 2010, Hejnice, June 21–25, and by Bjørn Jespersen at LRR 10, Ghent, September 20–22, 2010. This paper reproduces the bare semantic and logical bones required to validate quantifying into hyperintensional logic by means of TIL. For more details, see (Duží, Jespersen, & Materna, 2010, §5.3) and (Duží & Jespersen, n.d.). The research reported herein was supported by the Grant Agency of the Czech Republic, project No. 401/10/0792, "Temporal aspects of knowledge and information".

However, where ∂ is a dummy operator governing attitude contexts, this inference is controversial:

$$\frac{\partial\,Fa}{\exists x\,\partial\,(Fx)}$$

Its validity hinges on two factors absent in the initial case. First, whether in $\exists y(\partial\ldots y\ldots)$ the existential quantifier governing the context $(\partial\ldots y\ldots)$ succeeds in reaching across ∂ and binding any or all y occurring inside the scope of ∂. Second, whether a suitable range of quantification is available for y. The topic of quantifier-binding bears on the syntax of a given logical formalism, while the topic of quantificational range bears on the existential commitments of a given logical system. Quantifying *into* a hyperintensional context need not involve quantifying *over* hyperintensions. While TIL always validates the conclusion that there is a *hyperintension* c such that $(\partial\ldots c\ldots)$, TIL can also validate the stronger conclusion that there is an *intension* f (a function from possible worlds) such that $(\partial\ldots f_w\ldots)$. The conclusion then says that there is an f such that the value of f at the parameter (or context of evaluation) of the premise is ∂'ed to be an F. Our empirical parameters are pairs of worlds and times: $\langle w,t\rangle$. To keep our rule of quantifying into hyperintensional contexts as general as possible, it is formulated such that its conclusion quantifies over hyperintensions.

A dual constraint of TIL has impact on this rule and involves *properly partial functions*, which are undefined for some or all of their arguments, and *improper hyperintensions*, which fail to produce a product. Improperness basically arises from the procedure of applying a partial function f to an argument A, such that f returns no value at A. The procedure of functional application induces an extensional context. Thus when specifying various rules of quantifying-in, one must check whether particular constituent hyperintensions occurring extensionally are improper. If none is, the given rule of quantifying-in is valid. This paper builds especially upon (Tichý, 1986), which it enhances by extending it to hyperintensional attitudes, and (Materna, 1997), which it corrects and simplifies.

2 Semantic foundations

TIL is an extensional logic of hyperintensions, thus heeding the principles of compositionality and transparency of sense and denotation even for terms and expressions embedded in a hyperintensional context. Valid substitution inside hyperintensional contexts must be of pairs of synonymous terms, while co-denotation suffices for intensional contexts, where intensions are individuated up to necessary equivalence, as in modal logic. Transparency in hyperintensional contexts requires identity of senses (pairs of synonymous words), while transparency in intensional contexts requires only equivalence of senses (pairs of co-denoting words). In the final analysis, what gets substituted are not words, but their meanings.

The semantics is obtained by universalizing Frege's reference-shifting semantics custom-made for 'indirect' contexts found in the margins of Frege's overall extensionalist semantics. Thus, pairs of terms co-denoting outside an attitude context remain co-denoting inside an attitude context and pairs of terms that are not co-denoting inside an attitude context do not become co-denoting outside an attitude context. Our neo-Fregean semantic schema, which applies to all contexts, is this:

$$Expression \xxRightarrow{\text{expresses}} Construction \xrightarrow{\text{constructs}} Denotation$$
$$\underbrace{\hspace{7cm}}_{\text{denotes}}$$

E.g., definite descriptions that only contingently describe the same individual never qualify as co-denoting. Although Quine's 'the man in the brown hat' and 'the man on the beach' happen to co-refer to Bernard J Ortcutt, they do not co-denote him. Rather they denote, in every context, two distinct individual roles (modeled in TIL as functions from possible worlds to partial functions from times to individuals). There are worlds and times at which these two roles are co-occupied, e.g. by Ortcutt, but this empirical fact has no bearing on the semantic properties of these two definite descriptions. Their semantic properties concern instead whether they are synonymous (hence co-denoting) or merely co-denoting. Our strategy is to raise the bar somewhat for what qualifies as identity and equivalence of senses of words and apply substitution of identicals and equivalents to those (fewer) pairs of words that do pass muster. The opposite, and common, strategy is to maintain a lower bar, which, however, generates referential opacity and the inapplicability of substitution and quantification in various modal and attitudinal contexts.

The syntax of TIL is Church's (higher-order) typed λ-calculus, but with the all-important difference that the syntax has been assigned a procedural (as opposed to denotational) semantics, according to which a linguistic sense is an abstract procedure detailing how to arrive at an object of a particular logical type. TIL *constructions* are such procedures. Thus, *abstraction* transforms into the molecular procedure of forming a function, *application* into the molecular procedure of applying a function to an argument, and *variables* into atomic procedures for arriving at their assigned values. Constructions are linguistic senses, as well as modes of presentation of objects, and are our hyperintensions. Constructions are neither formulae (though encoded in an artificial language) nor functions (whereas Church thinks of senses as functions from senses to senses). Rather, technically speaking, some constructions are abstract modes of presentation of functions, including 0-place functions such as individuals and truth-values, and the rest are modes of presentation of other constructions. With constructions of constructions, constructions of functions, functions, and functional values in our stratified ontology, we need to keep track of the traffic between multiple logical strata. The *ramified type hierarchy* does just that. Constructions may themselves figure as functional arguments or values. Certain constructions, *qua* objects of attitudes, figure as functional arguments of the binary relations-in-intension that, relative to a world/time pair of evaluation, take an agent and an attitude relatum to a truth-value. Moreover, since constructions are being quantified into, we need constructions of one order higher than those being quantified into.

With both hyperintensions and possible-world intensions in its ontology, TIL has no trouble assigning either hyperintensions or intensions to variables as their values. However, the technical challenge of making x occurring in the context $\partial(\ldots x \ldots)$ bindable by $\exists$ requires two (occasionally three) interrelated, non-standard devices. The first is *Trivialization*, which is an atomic construction, whose only constituent is itself. The second is the function *Sub* (for 'substitution'). (The third is the function *Tr* (for 'Trivialization'), which takes an object to its Trivialization.) We say that Trivialization is *used* to *mention* other constructions. The point of mentioning a construction is to make it, rather than what it presents, a functional argument.

TIL construes ∂ as a binary relation-in-intension between an agent entertaining an attitude and a construction in its capacity as attitude relatum. Relations, in turn, are construed as functions, such that ∂ is construed as a function from worlds to a function from times to a function from pairs of agents (individuals) and hyperpropositions to truth-values: given a world/time pair, it is either true or false that a particular agent has a particular attitude to a particular hyperproposition. In order for the relevant construction to figure as the second argument of the relevant attitude relation, it itself needs to be mentioned. If a given hyperproposition is not mentioned but used, the resulting relatum is either a truth-value (which is unsuitable as attitude relatum) or a possible-world proposition, thus reinstalling the logic of attitudes known from modal logic.

For a construction to be mentioned is (in this paper) for it to be Trivialized. Therefore, in brief, a hyperpropositional attitude context is one in which the second argument of the attitude relation is a construction that constructs either a truth-value or a possible-world proposition (a function from worlds to a partial function from times to truth-values). The construction constructs a truth-value when the agent has a mathematical or logical attitude like being convinced that second-order logic is complete. This is so because empirical vicissitudes like modal or temporal variability do not impinge upon the truth-value of mathematical or logical propositions (which are constructions of truth-values). The construction constructs a possible-world proposition when the agent has an empirical attitude like expecting that the Evening Star is a star. This is so because modal and temporal variability does impinge upon the truth-value, at a given world/time pair of evaluation, of empirical propositions. Whether the construction be of a truth-value or of a possible-world proposition, the logical problem is that every construction, including variables, occurring inside the scope of a Trivialization occurs mentioned ('quoted') and, therefore, is not amenable to logical manipulation, such as $\exists$-binding. *Sub*, for its part, is a function that takes three constructions to a fourth construction. The first one is the construction to be substituted for the second one into the third one, and the fourth one is the resulting construction (see Tichý, 1988, p. 75).

3   Logical foundations

In this section we set out the definitions of *first-order types* (regimented by a simple type theory), *constructions*, and *higher-order types* (regimented by a ramified type hierarchy), which taken together form the nucleus of TIL, accompanied by some auxiliary definitions.

The type of first-order object includes all objects that are not constructions, such as the standard objects of individuals, truth-values, sets (see below for a restriction), as well as functions defined on possible worlds. *Sets* are always characteristic functions and insofar extensional entities, but the domain of a set may be typed over higher-order objects, in which case the relevant set is itself a higher-order object. Similarly for other functions, including *relations*, with domain or range in constructions. Whenever constructions are involved, we find ourselves in the ramified type hierarchy.

Definition 1 (*types of order 1*)   Let B be a *base*, where a base is a collection of pair-wise disjoint, non-empty sets. Then:

   (i) Every member of B is an elementary *type of order 1 over B*.

   (ii) Let $\alpha, \beta_1, \ldots, \beta_m$ ($m > 0$) be types of order 1 over B. Then the collection $(\alpha\beta_1 \ldots \beta_m)$ of all m-ary partial mappings from $\beta_1 \times \cdots \times \beta_m$ into α is a functional *type of order 1 over B*.

   (iii) Nothing is a *type of order 1 over B* unless it so follows from (i) and (ii).

Remark 1   For the purposes of natural-language analysis, we are currently assuming the following base of ground types, which is part of the ontological commitments of TIL:
   o the set of truth-values $\{\,T, F\,\}$;
   ι the set of individuals (the universe of discourse);
   τ the set of real numbers (doubling as discrete times);
   ω the set of logically possible worlds (the logical space).

Definition 2 (*construction*)

   (i) The variable x is a *construction* that v-constructs an object O of the respective type dependently on a valuation v.

(ii) *Trivialization*: Where X is an object whatsoever (an extension, an intension or a construction), ^0X is the construction *Trivialization*. It constructs X without any change.

(iii) The *Composition* $[X\,Y_1 \ldots Y_m]$ is the following *construction*. If X v-constructs a function f of type $(\alpha\beta_1 \ldots \beta_m)$, and Y_1, $\ldots$, Y_m v-construct entities B_1, $\ldots$, B_m of types β_1, $\ldots$, β_m, respectively, then the *Composition* $[X\,Y_1 \ldots Y_m]$ v-constructs the value (an entity, if any, of type α) of f on the tuple argument $\langle B_1, \ldots, B_m \rangle$. Otherwise the *Composition* $[X\,Y_1 \ldots Y_m]$ does not v-construct anything and so is v-improper.

(iv) The Closure $[\lambda x_1 \ldots x_m\, Y]$ is the following *construction*. Let x_1, x_2, $\ldots$, x_m be pair-wise distinct variables v-constructing entities of types β_1, $\ldots$, β_m and Y a construction v-constructing an α-entity. Then $[\lambda x_1 \ldots x_m\, Y]$ is the construction λ-*Closure* (or *Closure*). It v-constructs the function f of the type $(\alpha\beta_1 \ldots \beta_m)$. Let $v(B_1/x_1, \ldots, B_m/x_m)$ be a valuation identical with v at least up to assigning objects B_1/β_1, $\ldots$, B_m/β_m to variables x_1, $\ldots$, x_m. If Y is $v(B_1/x_1, \ldots, B_m/x_m)$-improper (see iii), then f is undefined on $\langle B_1, \ldots, B_m \rangle$. Otherwise the value of f on $\langle B_1, \ldots, B_m \rangle$ is the α-entity $v(B_1/x_1, \ldots, B_m/x_m)$-constructed by Y.

(v) The *Single Execution* ^1X is the *construction* that v-constructs the entity v-constructed by X. Otherwise ^1X is v-improper.

(vi) The *Double Execution* ^2X is the following *construction*. If X v-construct a construction Y and Y v-construct an entity Z, then ^2X v-constructs Z. Otherwise ^2X is v-improper.

(vii) Nothing is a *construction*, unless it so follows from (i) through (vi).

Definition 3 (*higher-order types*)

$\mathbf{T}_1$ (*types of order 1*). See Definition 1.

$\mathbf{C}_n$ (*constructions of order n*).

       i) Let x be a variable ranging over a type of order n. Then x is a *construction of order n over B*.

ii) Let X be a member of a type of order n. Then ^0X, ^1X, ^2X are *constructions of order n over B*.

iii) Let X, X_1, ..., X_m $(m > 0)$ be constructions of order n over B. Then $[X\,X_1 \ldots X_m]$ is a *construction of order n over B*.

iv) Let x_1, ..., x_m, X $(m > 0)$ be constructions of order n over B. Then $[\lambda x_1 \ldots x_m\,X]$ is a *construction of order n over B*.

v) Nothing is a *construction of order n over B* unless it so follows from C_n (i)-(iv).

$\mathbf{T}_{n+1}$ *(types of order $n+1$)*. Let $*_n$ be the collection of all constructions of order n over B. Then

i) $*_n$ and every type of order n are *types of order $n+1$*.

ii) If $0 < m$ and α, β_1, ..., β_m are types of order $n+1$ over B, then $(\alpha\,\beta_1 \ldots \beta_m)$ is a *type of order $n+1$ over B*.

iii) Nothing is a *type of order $n+1$ over B* unless it so follows from T_{n+1} (i) and (ii).

Definition 4 (*construction mentioned vs. used as a constituent*)
Let C be a construction and D a subconstruction of C.

(i) If D is identical to C (i.e., $^0C = {}^0D$) then the occurrence of D is *used as a constituent* of C.

(ii) If C is identical to $[X_1\,X_2 \ldots X_m]$ and D is identical to one of the constructions X_1, X_2, ..., X_m, then the occurrence of D is *used as a constituent* of C.

(iii) If C is identical to $[\lambda\,x_1 \ldots x_m\,X]$ and D is identical to X, then the occurrence of D is *used as a constituent* of C.

(iv) If C is identical to ^1X and D is identical to X, then the occurrence of D is *used as a constituent* of C.

(v) If C is identical to ^2X and D is identical to X, or ^0D occurs as a constituent of X and this occurrence of D occurs as a constituent of Y v-constructed by X, then the occurrence of D is *used as a constituent* of C.

(vi) If an occurrence of a subconstruction D of C is not *used as a constituent* of C then the occurrence of D is *mentioned* in C.

(vii) No occurrence of a subconstruction D of C is *used/mentioned* in C unless it so follows from (i)-(vi).

Definition 5 (*free and bound variables*) Let C be a construction with at least one occurrence of a variable ξ.

(i) Let C be ξ. Then the *occurrence of ξ in C is free*.

(ii) Let C be ^0X. Then every *occurrence of ξ in C is ^0bound* ('Trivialization bound').

(iii) Let C be $[X\,X_1\ldots X_n]$. Any *occurrence of ξ that is free*, ^0bound, λ-*bound* in one of X, X_1, $\ldots$, X_n is, respectively, *free*, ^0bound, λ-*bound* in C.

(iv) Let C be ^1X. Then any *occurrence of ξ that is free*, ^0bound, λ-*bound* in X is, respectively, *free*, ^0bound, λ-*bound* in C.

(v) Let C be ^2X. Then any *occurrence of ξ that is free*, ^0bound, λ-*bound* in a constituent ^0D of C and this occurrence of D is a constituent of X' v-constructed by X, then if the occurrence of ξ is free, λ-bound in D it is *free*, λ-*bound* in C. Otherwise, any other occurrence of ξ in C is ^0bound in C.

(vi) An occurrence of ξ is *free*, ^0bound, λ-*bound* in C only due to (i)-(v).

A construction with at least one occurrence of a free variable is an *open construction*. A construction without any free variables is a *closed construction*.

Definition 6 (*procedurally isomorphic constructions*)
Let C, D be constructions. Then C, D are α-*equivalent* iff they differ at most by deploying different λ-bound variables. C, D are η-*equivalent* iff one arises from the other by η-reduction or η-expansion. C, D are β_r-*equivalent* iff one arises from the other by β_r-reduction or β_r-expansion. C, D are *procedurally isomorphic* iff there are closed constructions C_1, $\ldots$, C_m, $m \geq 1$, such that $^0C = {}^0C_1$, $^0D = {}^0C_m$, and if $m > 1$ then all C_i, C_{i+1} ($1 \leq i < m$) are either α-, η- or β_r-equivalent.

Remark 2   Reduced β_r-conversion consists in substituting free variables for λ-bound variables of the same type.

Since merely co-referential expressions can be substituted *salva veritate* only in extensional context, merely co-denotational expressions in intensional and extensional contexts, and synonymous expressions in all contexts, we define:

Definition 7 (*synonymous, equivalent, co-referential expressions*)
Expressions E_1 and E_2 are *synonymous* if their meanings are procedurally isomorphic. Expressions E_1 and E_2 are *equivalent* (or *co-denoting*) if their meanings v-construct one and the same object for every valuation v. Finally, empirical expressions E_1 and E_2 are *co-referential* if their meanings construct intensions whose values are the same in the actual world at the present time.

Intensional entities are entities of type $(\beta\omega)$: mappings from possible worlds to an arbitrary type β. The type β is frequently the type of the *chronology* of α-objects, i.e., a mapping of type $((\alpha\tau)\omega)$, abbreviated as '$\alpha_{\tau\omega}$'. *Extensional entities* are entities of a type α where $\alpha \neq (\beta\omega)$ for any type β. Examples of frequently used *intensions* are: *propositions* of type $o_{\tau\omega}$, *properties of individuals* of type $(o\iota)_{\tau\omega}$, *binary relations-in-intension between individuals* of type $(o\iota\iota)_{\tau\omega}$, *individual offices/roles* of type $\iota_{\tau\omega}$.

Our *explicit intensionalization and temporalization* enables us to encode constructions of possible-world intensions, by means of terms for possible-world variables and times, directly in the logical syntax. Where w ranges over ω and t over τ, the following logical form (to be explained below) essentially characterizes the logical syntax of empirical language:

$$\lambda w \lambda t \, [\ldots w \ldots t \ldots]$$

Logical objects like *truth-functions* and *quantifiers* are extensional: $\wedge$ (conjunction), $\vee$ (disjunction) and $\supset$ (implication) are of type (ooo), and $\neg$ (negation) of type (oo). Quantifiers $\forall^\alpha$, $\exists^\alpha$ are type-theoretically polymorphous functions of type $(o(o\alpha))$, for an arbitrary type α, defined as follows. The universal quantifier $\forall^\alpha$ is a function that associates a class A of α-elements with **T** if A contains all elements of the type α, otherwise with **F**. The existential quantifier $\exists^\alpha$ is a function that associates a class A of α-elements with **T** if A is a non-empty class, otherwise with **F**. Below all type indications will be provided outside the formulae in order not to clutter the notation. Furthermore, 'X/α' means that an object X is (a member) of type α. '$X \to_v \alpha$' means that the type of the object v-constructed by X is α. We write '$X \to \alpha$' if what is v-constructed does not depend on a valuation v. Throughout, it holds that the variables $w \to_v \omega$ and $t \to_v \tau$. If $C \to_v \alpha_{\tau\omega}$ then the frequently used Composition $[[Cw]t]$, which is the intensional descent (a.k.a. extensionalization) of the α-intension v-constructed by C, will be encoded as 'C_{wt}'.

Only *procedurally isomorphic constructions*, as opposed to merely equivalent ones, are substitutable *salva veritate* in hyperintensional contexts. Church's Alternatives (0) and (1) leave room for additional Alternatives, like Alternative (1/2) and Alternative (2/3). The former includes α- and η-conversion while the latter adds a restricted β-conversion called β_r-conversion in Def. 6. For instance, we see no reason to differentiate between "b is believed by a to be happy" and "b has the property of being believed by a to be happy". The latter sentence expresses

$$\lambda w \lambda t \, [\lambda w' \lambda t' \lambda x \, [^0 Believe_{w't'} \, ^0 a \, \lambda w \lambda t \, [^0 Happy_{wt} \, x]]_{wt} \, ^0 b]$$

which is merely a β_r-expanded form of

$$\lambda w \lambda t \, [\lambda x \, [^0 Believe_{wt} \, ^0 a \, \lambda w \lambda t \, [^0 Happy_{wt} \, x]] \, ^0 b]$$

We currently prefer Alternative (2/3), to soak up those differences between β-transformations that lack natural-language counterparts, as our criterion of procedural isomorphism.

To compare intensional and hyperintensional attitudes, consider the sentence "a believes that the Evening Star is a planet" where 'the Evening Star' denotes the individual role of the brightest celestial body visible in the evening sky apart from the moon and 'to believe' is ambiguous between denoting an intensional or a hyperintensional doxastic relation, $B/(o\iota o_{\tau\omega})_{\tau\omega}$ or $B^*/(o\iota*_n)_{\tau\omega}$, respectively. Let $EvSt/\iota_{\tau\omega}$ be the role of Evening Star. Then the respective analyses are these:

$$\lambda w \lambda t \, [^0 B_{wt} \, ^0 a \, [\lambda w \lambda t \, [^0 Planet_{wt} \, ^0 EvSt_{wt}]]]$$
$$\lambda w \lambda t \, [^0 B^*_{wt} \, ^0 a \, ^0[\lambda w \lambda t \, [^0 Planet_{wt} \, ^0 EvSt_{wt}]]]$$

The difference is that B is a relation-in-intension between an agent a and what the Closure $\lambda w \lambda t \, [^0 Planet_{wt} \, ^0 EvSt_{wt}]$ constructs, namely a possible-world *proposition*, whereas B^* is a relation-in-intension between a and the Closure $\lambda w \lambda t [^0Planet_{wt} \, ^0EvSt_{wt}]$ itself, i.e. a *hyperproposition* (propositional construction), as encoded by the Trivialization of this Closure. In the idiom of procedural semantics, B^* relates a to the very procedure of constructing a proposition while B relates a to the product of this procedure.

4   The rule of quantifying-in

Quantifying-in comes easy when a's attitude is intensional:

$$\lambda w \lambda t \, [^0\exists \lambda f \, [^0B_{wt} \, ^0a \, [\lambda w \lambda t \, [^0Planet_{wt} \, f_{wt}]]]]$$

Types. $f/*_1 \to_v \iota_{\tau\omega}; \, \exists/(o(o\iota_{\tau\omega}))$
Gloss: "At $\langle w, t \rangle$, there is an individual role f such that a believes that its occupant (i.e., f_{wt}) is a planet."

Proof. At all $\langle w, t \rangle$-pairs the following steps are truth-preserving:

(1)	$[^0B_{wt} \, ^0a \, [\lambda w \lambda t \, [^0Planet_{wt} \, ^0EvSt_{wt}]]]$	assumption
(2)	^0EvSt is not improper	Def. 2
(3)	$[^0B_{wt} \, ^0a \, [\lambda w \lambda t \, [^0Planet_{wt} \, f_{wt}]]]$	$v(EvSt/f)$-constructs **T**, 1
(4)	$\lambda f \, [^0B_{wt} \, ^0a \, [\lambda w \lambda t \, [^0Planet_{wt} \, f_{wt}]]]$	v-constructs a non-empty class
(5)	$[^0\exists \lambda f \, [^0B_{wt} \, ^0a \, [\lambda w \lambda t \, [^0Planet_{wt} \, f_{wt}]]]]$	EG, 4

$\square$

A naive attempt to quantify into hyperintensional contexts along similar lines founders on a major technical complication:

$$\frac{\lambda w \lambda t \, [^0B^*_{wt} \, ^0a \, ^0[\lambda w \lambda t \, [^0Planet_{wt} \, ^0EvSt_{wt}]]]}{\lambda w \lambda t \, [^0\exists \lambda f \, [^0B^*_{wt} \, ^0a \, ^0[\lambda w \lambda t \, [^0Planet_{wt} \, f_{wt}]]]]}$$

Why is the conclusion no good?

The occurrence of f in $^0[\lambda w \lambda t \, [^0Planet_{wt} \, f_{wt}]]$—notice the leftmost Trivialization—is ^{0}bound, because the variable f occurs within the hyperintensional context of $[\lambda w \lambda t \, [^0Planet_{wt} \, f_{wt}]]$. So f is *mentioned*, hence not available for manipulation. It is shielded from $\exists$ by the first ^{0} in the Trivialization $^0[\lambda w \lambda t \, [^0Planet_{wt} \, f_{wt}]]$. A linguistic parallel would be to attempt to quantify into a quotational context, where the quotation marks would have an analogous shielding effect.

The solution consists in untying the occurrence of f from its ^{0}binding. To this end we use a *substitution technique*. A valid argument is obtained by applying the function *Sub* and a variable ranging over *constructions* of individual roles: $c/*_2 \rightarrow_v *_1$; $^2c/*_3 \rightarrow_v \iota_{\tau\omega}$. *Pre-processing* of the constituent $^0[\lambda w \lambda t \, [^0Planet_{wt} \, f_{wt}]]$ (i.e., application of *Sub*) serves to make the variable free for manipulation. *Sub* is of the polymorphous type $(*_n *_n *_n *_n)$ and operates on constructions in this way. Let X, Y, Z be constructions of order n. Then *Sub* is a mapping which, when applied to $\langle X, Y, Z \rangle$, returns the construction that is the result of correctly substituting X for Y in Z.

Another polymorphous function that we sometimes need (see below) when applying this substitution method is $Tr/(*_n \, \alpha)$ defined as follows (see Tichý, 1988, p. 68). Let α be a type of order n, o an object of type α. Then Tr is a function which, when applied to o, returns the Trivialization of o. There is an essential difference between using the construction Trivialization and applying the function Tr. For instance, whereas ^{03} constructs the number 3, the Composition $[^0Tr \, ^03]$ constructs the construction ^{03}. Whereas the Trivialization ^0x ^{0}binds the variable x and constructs just x, the variable x is free in the Composition $[^0Tr \, x]$, which v-constructs the Trivialization of the number that v assigns to x. For instance, $[^0Tr \, x] \, v(2/x)$-constructs the construction ^{02}.

For illustration, we analyze three different arguments that share the same premise, but have different conclusions. Two of them are invalid, while the valid argument forms the nucleus of our solution to the problem of quantifying into hyperpropositional attitudes *de dicto*.

Types: $B^*/(o\iota*_1)_{\tau\omega}$; a/ι; $Planet/(o\iota)_{\tau\omega}$; $ES/*_n \rightarrow \iota_{\tau\omega}$: a construction of the role $EvSt$; $f/*_1 \rightarrow_v \iota_{\tau\omega}$; $\exists/(o(o\iota_{\tau\omega}))$; $\exists^*/(o(o*_n))$; $c/*_2 \rightarrow_v *_1$; $^2c/*_3 \rightarrow_v \iota_{\tau\omega}$.

$$\frac{\lambda w \lambda t \, [^0B^*_{wt} \, ^0a \, ^0[\lambda w \lambda t \, [^0Planet_{wt} \, ES_{wt}]]]}{\lambda w \lambda t \, [^0\exists \lambda f \, [^0B^*_{wt} \, ^0a \, [^0Sub \, [^0Tr \, f] \, ^0f \, ^0[\lambda w \lambda t \, [^0Planet_{wt} \, f_{wt}]]]]]} \qquad (1)$$

$$\frac{\lambda w \lambda t \, [^0B^*_{wt} \, ^0a \, ^0[\lambda w \lambda t \, [^0Planet_{wt} \, ES_{wt}]]]}{\lambda w \lambda t \, [^0\exists^* \lambda c \, [^0B^*_{wt} \, ^0a \, [^0Sub \, c \, ^0c \, ^0[\lambda w \lambda t \, [^0Planet_{wt} \, ^2c_{wt}]]]]]} \qquad (2)$$

$$\frac{\lambda w \lambda t \, [^0B^*_{wt} \, ^0a \, ^0[\lambda w \lambda t \, [^0Planet_{wt} \, ES_{wt}]]]}{\lambda w \lambda t \, [^0\square^* \lambda c \, [^0B^*_{wt} \, ^0a \, [^0Sub \; c \, ^0c \, ^0[\lambda w \lambda t \, [^0Planet_{wt} \, c_{wt}]]]]]} \tag{3}$$

Argument (1) is invalid. To see why, suppose that the construction *ES* in the premise is the meaning of 'the brightest celestial body in the evening sky apart from the moon'. Although the class v-constructed by

$$\lambda f [^0B^*_{wt} \, ^0a \, [^0Sub \, [^0Tr \; f] \, ^0f \, ^0[\lambda w \lambda t \, [^0Planet_{wt} \, f_{wt}]]]]$$

is not empty, it is not guaranteed to be so by the premise. The substitution of the Trivialization of the role of Evening Star results in the Closure

$$[\lambda w \lambda t \, [^0Planet_{wt} \, ^0EvSt_{wt}]].$$

Although the Closures

$$[\lambda w \lambda t \, [^0Planet_{wt} \, ES_{wt}]]$$

and

$$[\lambda w \lambda t \, [^0Planet_{wt} \, ^0EvSt_{wt}]]$$

are equivalent, they are not procedurally isomorphic. The above argument would be valid only if B^* were replaced by B to engender an *intensional* attitude instead. But, an important reason for hyperpropositional attitude reports is to respect the attributee's perspective on an empirical situation, so a's attitude must be reproduced faithfully in order to obtain a valid inference. Therefore, the role of Evening Star must be conceptualized in the same manner both in the premise and the conclusion.

Argument (2) is also invalid. Validity requires the additional premise that a believes* that $^{20}C = C$, for any construction C, which requires a to recognize, for any instance of ^{20}C and C, that these two different constructions are equivalent. As soon as a's attitude is merely intensional, the equivalence $^{20}C = C$ suffices for validity. For illustration, assume again that a believes* that the Evening Star is a planet. Then there is an assignment to c (to wit, the Trivialization ^0EvSt such that substitution yields this Closure:

$$[\lambda w \lambda t \, [^0Planet_{wt} \, ^2[^0[^0EvSt]]_{wt}]]$$

Of course, $^2[^0[^0EvSt]]$ is equivalent to $[^0EvSt]$, both constructing the role of Evening Star. But, again, we must reproduce faithfully a's hyperintensional attitude in order to obtain a valid inference: the appropriate relatum is

$$[\lambda w \lambda t \, [^0Planet_{wt} \, [^0EvSt_{wt}]],$$

not

$$[\lambda w \lambda t \, [^0Planet_{wt} \, {}^2[^0[^0EvSt]]_{wt}]].$$

Argument (3) is the valid one. Here is the *proof* of its validity. $[^0Sub \, c \, {}^0c \, {}^0[\lambda w \lambda t \, [^0Planet_{wt} \, c_{wt}]]]$ v-constructs a Closure dependently on the valuation of c; the first occurrence of c is free here, and

- at $\langle w, t \rangle$ the Composition
  $[^0Sub \, c \, {}^0c \, {}^0[\lambda w \lambda t \, [^0Planet_{wt} \, c_{wt}]]]$ $v(ES/c)$-constructs the Closure $[\lambda w \lambda t \, [^0Planet_{wt} \, ES_{wt}]]$

- at $\langle w, t \rangle$ the Composition
  $[^0B^*_{wt} \, {}^0a \, [^0Sub \, c \, {}^0c \, {}^0[\lambda w \lambda t \, [^0Planet_{wt} \, c_{wt}]]]]$ $v(ES/c)$-constructs $\mathbf{T}$

- at $\langle w, t \rangle$ the Composition
  $[^0\exists \lambda c \, [^0B^*_{wt} \, {}^0a \, [^0Sub \, c \, {}^0c \, {}^0[\lambda w \lambda t \, [^0Planet_{wt} \, c_{wt}]]]]]$ v-constructs $\mathbf{T}$ as well, because the class of constructions v-constructed by $\lambda c [^0B^*_{wt} \ldots c \ldots]$ is non-empty.

Note, however, that $[^0Planet_{wt} \, c_{wt}]$ comes with *wrong typing* and so is necessarily an *improper* Composition. The variable c ranges over constructions of order 1 rather than intensions, so c cannot be Composed with w and t (see Definition 2). Yet the improperness of a construction matters only if the construction is introduced as *used* in order to produce a product, at which it fails. If a construction is merely *mentioned* as an argument of a function, it occurs as an ordinary object that can be operated on, and its failure to produce a product is logically immaterial. In the present case $\lambda w \lambda t [^0Planet_{wt} \, c_{wt}]$ is an argument of Sub, and the Composition $[^0Sub \, c \, {}^0c \, {}^0[\lambda w \lambda t [^0Planet_{wt} \, c_{wt}]]]$ $v(ES/c)$-constructs the Closure $[\lambda w \lambda t \, [^0Planet_{wt} \, ES_{wt}]]$, which is a *proper* constituent of a's hyperintensional attitude.

What makes the wrong typing in $[^0Planet_{wt} \, c_{wt}]$ logically justifiable is that the evaluation of c is *lazy*. The benefit of lazy evaluation includes, inter alia, the avoidance of error conditions in the evaluation of compound expressions when employing functional programming languages, which TIL is deeply inspired by, especially those centred around *call-by-need*. Only under strict evaluation will the evaluation of a term containing a failing sub-term itself fail. Lazy evaluation does not evaluate function arguments, unless their values are required to evaluate the function call itself.

The general *rule for quantifying into hyperpropositional attitudes de dicto* underlying the validity of argument (3) above can now be spelt out.

Let X be a constituent of a propositional construction C. Then the following *rule* is valid:

$$\frac{[^0B^*_{wt}\,^0a\,^0C(X/c)]}{[^0\exists^*\lambda c\,[^0B^*_{wt}\,^0a\,[^0Sub\ c\,^0c\,^0C(c)]]]}$$

Types. $X/*_n \rightarrow_v \alpha;\ c/*_{(n+1)} \rightarrow_v *_n;\ ^2c/*_{(n+2)} \rightarrow_v \alpha;\ a/\iota;\ \exists^*/(o(o*_n));\ B^*/(o\iota*_n)_{\tau\omega}$.

Proof. Let $[^0B^*_{wt}\,^0a\,^0C(X/c)]$ v-construct **T**. Then $[^0Sub\ c\,^0c\,^0C(c)]$ $v(X/c)$-constructs $C(X/c)$ and $[^0B^*_{wt}\,^0a\,[^0Sub\ c\,^0c\,C(c)]]$ $v(X/c)$-constructs **T**. Thus the class of constructions $\lambda c\,[^0B^*_{wt}\,^0a\,[^0Sub\ c\,^0c\,^0C(c)]]$ is non-empty and $[^0\exists^*\lambda c\,[^0B^*_{wt}\,^0a\,[^0Sub\ c\,^0c\,^0C(c)]]]$ v-constructs **T** as well. There is a valuation of the variable c, namely the construction X that is procedurally isomorphic with itself. Thus the result of the substitution of X for the variable c into the hyperintensional context embedded in $[^0B^*_{wt}\,^0a\,\ldots]$, namely $C(c)$, is the construction $C(X/c)$. Hence the conclusion must be true, so the argument must be valid.   □

5   Conclusion

Quantifying into hyperintensional contexts requires an extensional logic of hyperintensions. Much non-trivial footwork is required to lay out such a large-scale logical semantics. Once this is done, though, quantifying into hyperintensional contexts turns out to be as trivially valid as quantifying into 'extensional' contexts. However, since our logic comes with strong typing, quantifying into hyperintensional contexts introduces a technical complication absent in quantifying into intensional and extensional contexts. The complication is that the quantifier cannot bind any variables inside the hyperintensional context, thus rendering quantifying-in impossible. The solution consists in applying a substitution technique that makes the variables amenable to binding. In this manner quantifying into a hyperintensional context is rendered valid while respecting compositionality and transparency.

References

Duží, M., & Jespersen, B. (n.d.). Transparent quantification into hyperpropositional contexts de dicto.
(In submission)

Duží, M., Jespersen, B., & Materna, P. (2010). *Procedural Semantics for Hyperintensional Logic: Foundations and Applications of Transparent Intensional Logic* (Vol. 17). Dordrecht: Springer.

Materna, P. (1997). Rules of existential quantification into "intensional contexts". *Studia Logica*, *57*, 331–343.

Tichý, P. (1986). Indiscernibility of identicals. *Studia Logica*, *45*, 251–273. (Reprinted in Tichý, 2004, pp. 647–672)

Tichý, P. (1988). *The Foundations of Frege's Logic*. Berlin: De-Gruyter.

Tichý, P. (2004). *Pavel Tichý's Collected Papers in Logic and Philosophy*. Dunedin; Prague: University of Otago Press; Filosophia.

Marie Duží* and Bjørn Jespersen*,†
* Department of Computer Science,
VSB-Technical University Ostrava, 17. listopadu 15, 708 33 Ostrava
† Institute of Philosophy,
Czech Academy of Sciences, Jilská 1, 110 00 Praha 1

e-mail: `marie.duzi@gmail.com`
e-mail: `bjorn.jespersen@gmail.com`

The Actual Future in Formal Semantics

Andrea Iacona

Abstract

According to a view that may be called *actualism*, future contingents are either true or false. Their truth or falsity depends on what happens in one among the many futures that are possible, namely, the actual future. This paper addresses the question of how actualism can be accommodated in a formal system, and suggests that three shared assumptions might hinder the way to an answer: the first is that a tree structure is needed, the second is that tenses are to be analyzed in terms of operators, the third is that actuality must be formally represented.

1

The thought that underlies actualism is easy to grasp. Suppose that the following sentence is uttered now:

(1) It will rain.

A natural way to state the condition at which (1) is true is to say that (1) is true if and only if it will in fact rain. Since 'in fact' indicates actuality, this is to say that (1) is true if and only if it will actually rain. Thus, (1) is either true or false at any time t. For either it actually rains at some time later than t or it doesn't. In this respect, there is no difference between (1) and the following sentences:

(2) It rains.

(3) It rained.

Either it actually rains at t or it doesn't, and the same goes for any time earlier than t. Let a *history* be a possible course of events that stretches all the way back into the past and all the way forward into the future, and assume that there is a plurality of histories, one for each possible continuation of the present state of affairs. Actualism entails that the truth-value of (1), just like that of (2) or (3), depends on what happens in one in particular of these histories, the actual history.[1]

The question of how this thought can be rendered formally has been addressed by friends and enemies of actualism. The options that have been considered so far rest on three assumptions. The first is that a tree structure is needed, that is, a structure formed by a set M of moments and a non-linear order $<$ on M. In such a structure, histories are maximal chains of moments, that is, maximal linearly ordered subsets of M. As figure 1 shows, three moments m_0, m_1 and m_2 in M may be such that $m_0 < m_1$ and $m_0 < m_2$ but it is not the case that $m_1 < m_2$. Thus, h_1 and h_2 may be different histories such that h_1 includes m_0 and m_1 while h_2 includes m_0 and m_2.

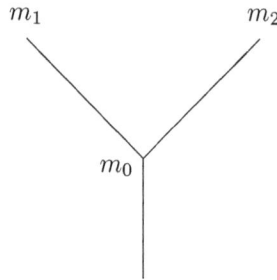

Figure 1: Tree

The second assumption is that tenses are to be analyzed in terms of operators. As Arthur Prior has shown, a language of tense logic is easily obtained by enriching a propositional language with temporal operators. If p stands for (2), then (1) is represented as Fp, where F stands for 'It will be the case that', and (3) is represented as Pp, where P stands for 'It was the case that'. In a tree structure, Fp and Pp can be evaluated at moments relative to histories that include

---

[1] Iacona (2011) spells out and defends this view.

those moments: Fp is true at m/h if and only if for some m' such that $m < m'$, p is true at m'/h; Pp is true at m/h if and only if for some m' such that $m' < m$, p is true at m'/h. Suppose for example that it rains at m_2 but not at m_1 or at any other moment in h_1. Then Fp is true at m_0/h_2 but false at m_0/h_1.[2]

The third assumption is that actuality must be formally represented, that is, something must be added to the structure to make sense of the idea that truth is a matter of actuality. In figure 1 we see that m_1 and m_2 are equally possible at m_0, but we do not see which of them is the state of affairs that will in fact obtain. In other words, the structure doesn't tell us whether h_1 or h_2 is the actual course of events. But actualism, one might think, requires that reference is made to a distinguished history as the actual history. The history in question is what Nuel Belnap and others call the *Thin Red Line*.[3]

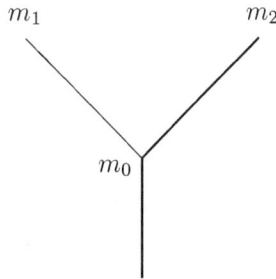

Figure 2: Tree with Thin Red Line

2

The simplest way to construct a semantics for actualism in accordance with these three assumptions is to take a tree structure and mark one of the histories in the structure as the actual history. Let $h_@$ be the

---

[2]The analysis of tenses in terms of operators is introduced in (Prior, 1957). A semantics of the kind outlined is presented in (Prior, 1967, p. 126 ff.) and (Thomason, 1970).

[3](Belnap & Green, 1994, pp. 379–381) and (Belnap, Perloff, & Xu, 2001, pp. 160–170).

history in question. Truth at m can be defined as follows for a formula α:

Definition 1 α is true at m iff α is true at $m/h_@$.

Suppose that $h_@ = h_2$, as in figure 2, and that Fp is true at m_0/h_2. Then Fp will be true at m_0, no matter whether it is true or false at m_0/h_1.[4]

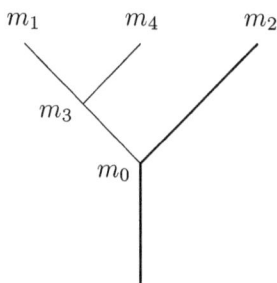

Figure 3: Counterfactual predictions

One problem that has been raised in connection with this proposal concerns the evaluation of formulas at moments that do not belong to $h_@$. Let m_3 and m_4 be as in figure 3, and let h_3 be a history that includes m_3 and m_4. Let q stand for 'It snows', and suppose that it snows only at m_1. Fq turns out false at m_3, because m_3 does not belong to $h_@$. But this seems wrong. From the point of view of m_2, one might say what follows, referring to a moment like m_1:

(4) Had things gone that way, one would have made a true prediction by uttering 'It will snow' some time before.

There are cases in which it is plausible to regard a counterfactual prediction as true. In those cases, the intuitive evaluation of the prediction does not depend on what happens in the actual history, but rather on what happens in some other history that *would* then be actual. In the example considered, the alternative that seems relevant

---

[4]Belnap et al. (2001, pp. 162–164) and MacFarlane (2003, p. 325) consider a semantics of this kind.

to the truth of Fq at m_3 is whether things go as in h_1 or as in h_3. This is why Fq seems true on the hypothesis that h_1 is actual.[5]

The problem of counterfactual predictions arises because the actual history is fixed in the structure once and for all, hence it does not vary with the moment of evaluation. So the most reasonable alternative to the semantics just sketched is to assume that there is a function from moments to histories that include them, that is, a function that assigns to each moment its own actual history. Let $h_@(m)$ be the history assigned to m. Then definition 1 may be rephrased as follows:

Definition 2 α is true at m iff α is true at $m/h_@(m)$

The hypothesis that at m_3 thing go as in h_1 is now represented as the hypothesis that $h_@(m_3) = h_1$, as in figure 4. So Fq turns out true at m_3 on that hypothesis.[6]

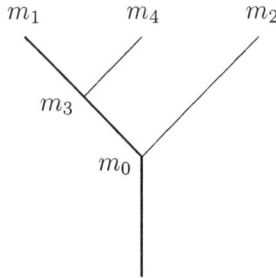

Figure 4: The relativization strategy

However, this proposal is affected by a different problem. First of all, note that the following principle must be rejected

(5) If $m < m'$ then $h_@(m) = h_@(m')$.

Suppose that (5) holds. If m_0, m_1 and m_2 are as in figure 1, from (5) we get that $h_@(m_1) = h_@(m_2)$. Since $m_1 \in h_@(m_1)$ and $m_2 \in$

---

[5] Belnap et al. (2001, pp. 162–163) draws attention to this problem.

[6] Braüner, Hasle, and Øhrstrøm (1998) and Øhrstrøm (2009) propose a semantics of this kind. A different version, suggested by McKim and Davis (1976), is that in which the value of the function for m is a chain that starts with m, rather than a whole history. Belnap et al. (2001, pp. 165–168) considers both options.

$h_@(m_2)$, it follows that $m_1 \in h_@(m_2)$ and $m_2 \in h_@(m_2)$, which contradicts the assumption that $h_@(m_2)$ is a maximal chain. Secondly, note that if (5) is rejected, the following basic principle of tense logic must be rejected as well:

(6) If p then it was the case that it would be the case that p.

Suppose that $h_@(m_3) = h_1$, as in figure 4, and that for any m such that $m < m_3$, it is not the case that $m_3 \in h_@(m)$. If it snows only at m_1, Fq turns out false at any m such that $m < m_3$, and the same goes for FFq. So Fq is true at m_3, while $PFFq$ is false. As Belnap and others have observed, the failure of (6) is not a desirable result. Imagine that a coin is flipped at m_0, that heads is the outcome at m_1, even though tails can be the outcome at m_2. From the point of view of m_2, one would be forced to say something like "the coin has landed tails, but that is not what was going to happen at m_0: at m_0 it was going to land heads".[7]

3

§2 outlines two ways to construct a semantics for actualism, and considers two problems that have been raised in connection with them. Further variants and refinements might be discussed. But what matters for the purpose at hand are the three assumptions that underly such attempts. In this section it will be suggested that the assumptions in question are not to be taken for granted.

The main motivation for the first, that a tree structure is needed, goes as follows. Actualism is tenable only if it is consistent with indeterminism. But indeterminism entails *branching*, the view according to which two histories can share a temporal part. On that view, our past is such a part, as it is the trunk from which a plurality of possible futures radiate. Since a tree structure offers a formal representation of branching, actualism must be accommodated in a tree structure.[8]

However, it is questionable that indeterminism entails branching. Certainly, indeterminism may be understood in many ways, and pre-

---

[7] (Belnap & Green, 1994, p. 380) and (Belnap et al., 2001, pp. 166–167). Øhrstrøm (2009, pp. 29–30) suggests that the problem can be avoided if the definition of the function is suitably adjusted.

[8] This line of thought emerges clearly in (Belnap et al., 2001, pp. 133–141) and in (MacFarlane, 2003, pp. 323–324).

sumably on some of them branching holds by definition. But there is at least one plausible understanding of indeterminism on which it does not entail branching. Let a *state* be a way in which the universe can be at a time. If S is a state that obtains at t and S' is a state that obtains at t', S *determines* S' if and only if the fact that S obtains at t and the laws of nature entail that S' obtains at t'. Determinism is the claim that the state of the universe at any time is determined by its state at earlier times, indeterminism is the negation of that claim. Indeterminism so understood does not entail branching, as it simply requires that two histories can be in the same state at t, just as at any earlier time, but in different states at t', neither of which is determined by their state at t.

Indeterminism may equally be framed in terms of the view that David Lewis calls *divergence*. On that view histories do not overlap, although they can have qualitatively identical parts. Possible futures may be conceived as parts of possible worlds that are wholly distinct, rather than branches that depart from a common trunk. So at each time in each world there is a unique future, namely, the series of events that occur at later times in that world. A formal representation of divergence is offered by $T \times W$ structures. A $T \times W$ structure includes a set T of times, a linear order $<$ on T, and a set W of worlds. Formulas are evaluated at time-world pairs, and the operators F and P are interpreted as follows: Fp is true at (t, w) if and only if, for some t' such that $t < t'$, p is true at (t', w); Pp is true at (t, w) if and only if, for some t' such that $t' < t$, p is true at (t', w). Worlds may be represented as vertical lines that run parallel, while times may be represented as horizontal lines that cut across them. For example, the grid in figure 5 describes a situation of the same kind described by the tree in figure 1, if we suppose that it rains at (t_1, w_2) but not at (t_1, w_1).[9]

There is a clear difference between figure 5 and figure 1. While (t_1, w_1) amounts to m_1 and (t_1, w_2) amounts to m_2, no single point amounts to m_0, for w_1 and w_2 include different points at t_0. This difference, however, does not matter as far as indeterminism is concerned. For w_1 and w_2 are like h_1 and h_2 in the relevant sense, that is, they are in the same state up to t_0 but in different states at t_1, neither

---

[9]Lewis (1986, pp. 206–209) spells out the difference between branching and divergence. Iacona (2011) provides a more articulated discussion of the claim that indeterminism entails branching. Thomason (1984) outlines $T \times W$ semantics.

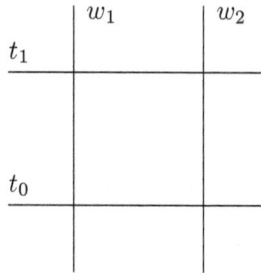

Figure 5: Grid

of which is determined by their state at t_0. The general point that
emerges from the contrast between tree structures and $T \times W$ structures may be stated as follows. As far as the formal representation of
indeterminism is concerned, all is needed is that truth is defined relative to worlds or histories, so that a formula that represents a sentence
such as (1) may take different values in different worlds or histories.
Any further issue concerning overlap is inessential.

The second assumption, that tenses are to be analyzed in terms
of operators, has always been matter of controversy. Much of the ink
that has been spilled on foundational issues that pertain to tense logic
concerns the plausibility of this assumption. A classical alternative to
the analysis in terms of operators is the analysis in terms of quantification over times. According to the latter, (1) is to be read as 'For
some t later than the present time, it rains at t', (2) is to be read as
'There is a time t such that t is the present time and it rains at t',
and (3) is to be read as 'For some t earlier than the present time, it
rains at t'.

A major argument in support of the analysis in terms of operators
goes as follows. The hypothesis that (2) involves a quantification
over times is implausible, for it makes the logical form of (2) more
complex than it appears. According to a version of the argument
due to Hans Kamp, the problem lies in the fact that no reference
to abstract entities such as times seems involved in (2). According
to another version due to Patrick Blackburn and Francois Recanati,
the problem lies in the departure from the "internal" perspective on
time, that is, the perspective we have as speakers situated inside the

temporal flow. Blackburn and Recanati argue that the present tense is not a tense like the past or the future. It is more primitive and, in a sense, temporally neutral.[10]

However, much in this line of reasoning depends on what kind of thing logical form is expected to be. If logical form is taken to be a property of sentences that is fixed by their truth-conditions independently of their surface grammar, it is conceivable that the logical form of (2) is more complex than it appears. In that case, it may be contended that no consideration concerning the apparent structure of (2) or the cognitive aspects pertaining to its use is able to undermine the hypothesis that the logical form of (2) is expressed by a quantification over times. Truth-conditions may not be detectable from surface grammar, and certainly transcend the "internal" perspective on time.

More generally, the question of how tenses are to be analyzed hinges on the thorny issue of how logical form is to be individuated. Although the latter issue is too big to be addressed here, for the purpose at hand it suffices to recognize that the analysis in terms of quantification over times is tenable on some understanding of logical form that is consistent with actualism. This means that nothing prevents us from thinking that actualism can be formalized without the help of tense logic.

The third assumption, that actuality must be formally represented, rests on the thought that in order to make sense of actualism, a formal definition of truth in terms of actuality must be provided. However, as the two problems considered in §1 show, it is hard to substantiate that thought. In particular, a definition of the kind envisaged seems unable to account for the fact that each history is actual from its own point of view. Both problems, in different ways, involve this shortcoming.

Note that the adoption of tree structures is not essential in this respect. Similar problems arise with $T \times W$ structures if truth is defined in terms of actuality. Let $w_@$ be the actual world in a $T \times W$ structure. A definition of truth at t in the spirit of definition 1 is the following:

Definition 3 α is true at t iff α is true at $(t, w_@)$.

In this case we get the same problem of counterfactual predictions. Let w_1 and w_2 be such that $w_2 = w_@$, as in figure 6, and suppose that

---

[10](Kamp, 1971, p. 231); (Blackburn, 1994, p. 83) and (Recanati, 2007, p. 70).

it snows only at (t_1, w_1). From the point of view of (t_1, w_2), it seems right to utter (4) referring to (t_1, w_1). But Fq is simply false at t_0, for it is false at (t_0, w_2). As in the case of definition 1, the trouble is that when we think about (t_0, w_1) and wonder whether Fq is true, the relevant question is what is actual from the point of view of w_1, not from that of w_2.[11]

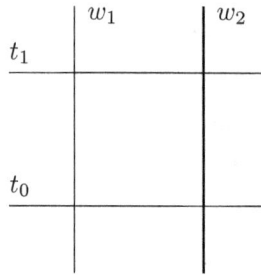

Figure 6: Grid with Thin Red Line

Again, no improvement is obtained by relativization. In order to have a definition in line with definition 2 it should be assumed either that actuality varies with the time of evaluation, or that it varies with the time-world pair of evaluation. But neither option is promising. In the first case, no real alternative to definition 3 is obtained. For a function from times to worlds gives us a unique actual world, namely, that defined by the function itself. In the second, the variation is easily defined as a function from time-world pairs to the worlds they include, that is, $w_@(t, w) = w$. Thus,

Definition 4 α is true at (t, w) iff α is true at $(t, w_@(t, w))$.

But in this case $(t, w_@(t, w)) = (t, w)$, so the definition boils down to a tautology. If actuality varies with the time-world pair of evaluation, reference to actuality becomes redundant, and the semantics becomes trivially equivalent to a $T \times W$ semantics without distinguished world.

The troubles that affect the attempts to provide a formal definition of truth in terms of actuality suggest that the crucial question to be

---

[11]A semantics of this kind is considered in (Kutschera, 1997, p. 248).

addressed is whether actuality must be formally represented. Actualism explains truth *simpliciter* as truth in the actual history. That is,

(7) A future contingent is true if and only if it is true in the actual history.

So the question is whether some *ad hoc* semantics is required by this explanation, that is, some semantics that does not simply define truth relative to a history. Contrary to the proposals considered so far, it may be claimed that no such semantics is required. Take (1). What (7) entails is that (1) is true if and only if it is true in the actual history. The right-hand side of this biconditional amounts to the conjunction of two conditions: (a) there is a history h in which (1) is true, and (b) h is the actual history. A formal semantics can provide a characterization of (a) to the extent that in some structure $\mathcal{A}$ in which h is represented, a formula α that stands for (1) is true relative to h. Once that characterization is given, (b) may be understood as a hypothesis about $\mathcal{A}$. That is, $\mathcal{A}$ gives us an account of the truth of (1) on the hypothesis that h is the actual history. (b) need not be expressed in the semantics, because it is a hypothesis *about* the semantics.

The plausibility of this way of understanding (7) turns out clear if we leave aside time and just think about modality. The semantics of a standard language of modal logic provides a definition of truth relative to a world. However, anyone agrees that such a definition is not to be taken as an analysis of truth *simpliciter*. The tacit assumption is that truth *simpliciter* involves something more than that, namely, actuality. Truth relative to a world offers a representation of truth *simpliciter* on the hypothesis that the world in question is the actual world.

In substance, actualism is consistent with a formal semantics in which the only definition of truth is relative to worlds or histories. All that actualism requires is that truth *simpliciter* does not reduce to the property so defined. To say that (1) is true is not the same thing as to say that it is true in some world, nor is the same thing as to say that it is true in all worlds. Actualism differs both from a view that rejects the idea of truth *simpliciter* and from one that identifies truth *simpliciter* with truth in all worlds. Yet the formal semantics can be exactly the same. The difference is not in the semantics, but rather in the way the semantics is understood.

Thus, it turns out that the three assumptions that have been taken for granted by friends and enemies of actualism are dispensable: actualism may be accommodated in a formal system without tree structures, without temporal operators, and without distinguished history. There is at least one way to go that is consistent with the rejection of these three assumptions. It may be argued that a suitable logic for actualism is the simplest quantified modal logic. But that is matter for another paper.

References

Belnap, N., & Green, M. (1994). Indeterminism and the Thin Red Line. In J. Tomberlin (Ed.), *Philosophical perspectives* (Vol. 8, pp. 365–388). Ridgeview.

Belnap, N., Perloff, M., & Xu, M. (2001). *Facing the Future*. Oxford University Press.

Blackburn, P. (1994). Tense, Temporal Reference and Tense Logic. *Journal of Semantics*, *11*, 83–101.

Braüner, T., Hasle, P., & Øhrstrøm, P. (1998). Ockhamistic Logics and True Futures of Counterfactual Moments. In I. Press (Ed.), *Proceedings of the fifth international workshop on temporal representation and reasoning* (pp. 132–139).

Iacona, A. (2011). Timeless Truth. In F. Correia & A. Iacona (Eds.), *Around the Tree: Semantic and Metaphysical Issues concerning Branching and the Open Future*. (forthcoming)

Kamp, H. (1971). Formal Properties of "now". *Theoria*, *37*, 227–73.

Kutschera, F. von. (1997). TxW Completeness. *Journal of Philosophical Logic*, *26*, 241–250.

Lewis, D. (1986). *On the Plurality of Worlds*. Blackwell.

MacFarlane, J. (2003). Future Contingents and Relative Truth. *Philosophical Quarterly*, *53*, 321-336.

McKim, V. R., & Davis, C. C. (1976). Temporal Modalities and the Future. *Notre Dame Journal of Formal Logic*, *17*, 233–238.

Øhrstrøm, P. (2009). In Defence of the Thin Red Line: a case for Ockhamism. *Humana Mente*, *8*, 17–32.

Prior, A. N. (1957). *Time and Modality*. Oxford University Press.

Prior, A. N. (1967). *Past, Present and Future*. Clarendon Press.
Recanati, F. (2007). *Perspectival Thought*. Oxford University Press.
Thomason, R. H. (1970). Indeterminist Time and Truth-Value Gaps. *Theoria*, *36*, 264-281.
Thomason, R. H. (1984). Combinations of Tense and Modality. In D. Gabbay & G. Guenthner (Eds.), *Handbook of philosophical logic* (Vol. 2, pp. 135–165). Reidel.

Andrea Iacona
Dipartimento di Storia e Metodologie Comparate
Universita dell'Aquila
Via dell'Industria, Bazzano (AQ), Italy
e-mail: `ai@cc.univaq.it`

Affirmation and Denial, Judgments of Composition and of Division

John T Kearns

Abstract

A statement is a true or false speech act performed with (what is normally) a declarative sentence. An assertion is an illocutionary act performed by making a statement and accepting the statement as being or representing what is the case. And a denial is an illocutionary act which rules out the assertion of a statement for objective reasons. In simple cases, we often deny a statement by using a negative expression to separate or divide a predicating act from a referring act. Denial is more fundamental than statement negation, and is the source of the significance of statement negation. Since it can never make sense to both accept a statement and rule out the acceptance of that statement, neither can it make any sense to accept a statement and its negation.

1 Language acts

When we speak, write, or think with words, we are performing *speech acts*, or *language acts*. Some language acts are performed with whole sentences, and some of these are true or false. I will understand a *statement* to be a language act that employs a whole sentence and that can appropriately be judged to be true or false; it is *true* if it "fits" the world, and *false* if it fails to fit. Some language acts performed with whole sentences are *illocutionary acts*, these are intentional acts, and are something like the minimal complete sentential acts provided for by linguistic conventions. Assertions, denials, promises, threats, and requests are illocutionary acts. An illocutionary act of a certain kind is said to have an *illocutionary force* which determines it to be an act of that kind.

As I have characterized statements, they are not illocutionary acts. But a statement can be performed with one or another illocutionary force, and will then constitute one or another kind of illocutionary act. For example, a statement can be asserted or denied. Informally, we might regard the statement as a component of such an illocutionary act, as the "content" of the act perhaps. More accurately, the statement is the illocutionary act abstractly conceived. However, not all statements are performed with illocutionary force. Anyway, the point of considering statements is to abstract away from considerations of illocutionary force. This is what makes them something less than complete acts.

Although it is common to regard all illocutionary acts as communicative acts which are performed by an agent and aimed at addressees, I am conceiving illocutionary acts more broadly. For some illocutionary acts such as directives and promises, an addressee is essential. You can't ask someone to pass the salt if there is no someone there. But as I am understanding assertions, for example, what is essential is that the speaker/writer/thinker produces (performs) a statement and accepts it as being or representing what is the case. A speaker can make an assertion by which she comes to accept the statement, or she can make an assertion to reaffirm her continued acceptance of the statement. Although a speaker can address her assertion to someone else, I count it as an assertion if she judges a statement to be the case when she is alone, without telling anyone else.

When a factual statement is accepted, it is accepted as being, or representing, things as they are. For an act of asserting/accepting a statement to be successful, the statement must be true, and the person who accepts the statement must have adequate grounds for accepting it. But in accepting a statement, a person is not at the same time characterizing the statement. In accepting a statement A, she is not accepting the statement "It is true that A"; neither is she accepting the statement "I know that A" or "I have good grounds for A."

Statements can be asserted/accepted or denied, but there is a sense in which statements are "designed" for being asserted or accepted. In a simple statement, where an expression is predicated of one or more objects, the predicate expression is associated with a criterion, and is truly predicated of an object or objects that satisfy this criterion.

The whole "idea" of a predicate expression is *that it be applied to objects which satisfy its criterion.* When a predicate is applied to objects which satisfy its criterion, we can say that it *fits* those objects. A predicate is, by design, intended to fit objects, much as a coat is intended, by design, to fit certain wearers. A person might apply a predicate to an object it doesn't fit, and might even do this deliberately, but an application of a predicate to an object is *fully successful* (perhaps we should say *intrinsically successful*) just in case the predicate fits the object.

Similarly, a statement constituted by predicating an expression of an object or objects is fully successful if its predicative component is fully successful—in that case the statement presents or represents things as they are. A fully successful statement *fits the world, or part of it.* More complex statements also fit or fail to fit the world, but their conditions for fitting are more complicated than for simple statements. Because statements are, by design, intended to fit the world, asserting or accepting is what "comes naturally" to statements—these acts endorse the statements for being successful.

A denial might be characterized as an act of rejecting a statement, but what is really rejected is the illocutionary act which the statement is designed to perform. In denying a statement, a speaker or language user is *ruling out* the assertion, the acceptance, of that statement. The assertion is ruled out, not because of ignorance on the part of the speaker, but because the statement is objectively unsuited for being asserted. In many natural languages, it is common to deny a simple statement by interfering with, or interrupting, the statement that is denied. For example, someone might use this sentence:

> Prague is not in Austria.

to rule out the acceptance of the statement that Prague is in Austria. In making this denial, the word 'not' does not function as the negation operator in systems of propositional logic. Instead, it marks the illocutionary force of denial by interrupting the predication of 'is in Austria' of Prague. The denial *divides* the act of referring to, or representing, Prague from the predicating act. Simple assertions accept statements *composed* by combining referring acts with predicative acts; simple denials *divide* those acts.

I understand denial to be more fundamental than, and logically prior to, statement negation. Denial can be used to introduce and (partially) characterize propositional negation. In systems of *illocu-tionary logic*, I use the following *illocutionary operators*, prefixed to sentences, to represent illocutionary acts:

⊢ — the sign of assertion
⊣ — the sign of denial

If A is (represents) a statement, then the assertion of A and the denial of A can be represented as follows:

$$\vdash A \qquad\qquad \dashv A$$

The illocutionary operators are *first-person* operators, they are supplied by the speaker (the language user) to make explicit the force of her illocutionary act. Then if we use '$\sim$' for statement negation, the inference principles shown here might be used to introduce and characterize statement negation:

$\sim$ *Introduction* $\sim$ *Elimination*

$$\frac{\dashv A}{\vdash \sim A} \qquad\qquad \frac{\vdash \sim A}{\dashv A}$$

Understood in terms of these principles, to apply the sign of negation to a statement is to characterize that statement as one that, for ob-jective reasons, can appropriately be denied. Since a true statement is one that "fits" the world, the objective reasons that make a statement false are that it fails to fit the world.

Given the understanding of assertion and denial that has just been sketched, it makes no sense, it is incoherent, to both assert and deny the same statement. Consider this analogical situation: A person can't, at one and the same time, both open and refuse to open the door. Because of this, it is irrational, it is senseless, for one person to direct someone else to both open and not open the door. The person issuing such a directive could not rationally intend for the addressee to carry out, to *implement*, the directive. Because it is essential to a directive that the speaker wants the addressee to carry out, or, at least, try to carry out what is directed, the speaker who tells someone else to both open and not open the door has not succeeded in issuing a directive.

A speaker can direct her addressee to open the door, then change her mind, and tell that addressee not to open the door after all. She can't successfully direct her addressee to do both. And a speaker can assert a statement, then change her mind, take back the assertion, and instead deny the statement. She can't significantly both assert and deny the statement. To rule out the assertion of a statement is not to say something *about* the statement, it is not to *characterize* the statement. Ruling out the assertion blocks the assertion for the speaker, and presents this blocking as being objectively appropriate for everyone. If statement negation is characterized by the principles above, it also makes no sense to accept both a statement and its negation.

2 Suppositions

In a genuine argument, a *speech-act argument*, that a person uses to either extend her own knowledge or to persuade someone else, the components are illocutionary acts, and not simply statements. A person begins by making assertions or denials, and concludes with an assertion or denial. But in addition to assertions and denials, another kind of illocutionary act performed with statements plays an important role in deductive arguments. We assert statements and we deny statements, but we also *suppose* statements to be the case. These are *factual* rather than counterfactual suppositions. To suppose a statement to be the case is to accept the statement temporarily, in order to explore the consequences of combining that statement with what we know, or with what we believe and disbelieve. The statements we accept temporarily *supplement* the statements we (fully) accept. If we suppose some statements, and combine these with assertions and denials which register our knowledge or belief, any conclusions we derive will have the status of suppositions, and I call them suppositions. The expression I use to indicate the force of supposing a statement to be the case is this: ∟

The act of supposing a statement to be the case is positive, and it has a negative counterpart, as assertion has a negative counterpart in denial. We can't temporarily block the assertion of a statement, because suppositional acts don't interfere with assertions and denials. The negative counterpart to supposition, which I will simply call *negative supposition*, (temporarily) blocks or impedes the temporary acceptance of a statement. The symbol I will use for negative supposition is the symbol often used for Intuitionist negation: ¬. This is obtained by rotating the symbol for (positive) supposition *180* degrees, as the symbol for denial is obtained from the symbol for assertion. An act of negatively supposing statement A ($\neg A$) is incoherent with an act of asserting A ($\vdash A$), but it is "outweighed" by the assertion. Given these two acts:

$$\underline{\vdash A \qquad \neg A}$$

the conclusion that is logically required is this: $\vdash A$; this move would cancel or discharge a hypothesis from which '$\neg A$' is derived—the hypothesis might be the negative supposition '$\neg A$' itself.

The negative supposition of A is incoherent with the positive supposition of A. But it isn't incorrect to make incoherent suppositions. We often do this as part of a strategy that leads to a move which discharges a hypothesis. However, when we have incoherent suppositions (or an incoherent combination of suppositions, assertions, and denials), there is an over-riding commitment to remove the incoherence. The negative supposition of a statement impedes the positive supposition of that statement in the sense of making the positive supposition incoherent.

We should distinguish negatively supposing a statement from temporarily *suspending* or *withdrawing* our support of assertions and denials we have made, which we might do in order to consider the consequences of a limited portion of our knowledge and belief. No illocutionary acts are required for doing this; we simply ignore, disregard, some of what we know, believe, or disbelieve. If we liked, we could introduce symbols for *declining* to perform illocutionary acts. For when we withdraw our acceptance of statement A, we aren't *denying* A, and we aren't negatively supposing A, we are declining to assert A. It is possible to investigate acts of declining to assert, acts of declining to deny, and even acts of declining to suppose a statement. But none of these are acts which commit us to assert, deny, or suppose anything. Acts of declining to assert, deny, or suppose have little interest for the study of deduction.

The principles $\sim$ *Introduction* and $\sim$ *Elimination* for assertion and denial have these counterparts for supposition:

$\sim$ *Introduction* $\qquad\qquad$ $\sim$ *Elimination*

$$\frac{\neg A}{\llcorner\sim A} \qquad\qquad\qquad \frac{\llcorner\sim A}{\neg A}$$

The principles for assertion, denial, and positive and negative supposition, provide a reasonably complete analysis of the commitments
associated with negation. They indicate how we intend to understand
negation.

3   The logic of statement negation

For a single statement A, the acts of asserting A and denying A are
logically incoherent (with one another). A person might, in fact, be
committed by her beliefs and disbeliefs to perform such incoherent
acts. But this is a sign of error. Although there is a sense in which
a person whose assertions and denials are incoherent is committed
to accept every statement and to deny every statement, this is not
a serious commitment. The over-riding commitment to achieve coherence "outranks" the commitment to accept and deny everything.
The person whose assertions and denials are incoherent is rationally
committed to eliminate this incoherence, even though it may not be
easy for her to determine just how to eliminate it. It would be irrational for her to realize that her beliefs were incoherent, and to think
that this didn't matter. It would be more than irrational, it would
simply be crazy, if she both was untroubled by the incoherence, and,
further, proceeded to infer arbitrary assertions and denials from her
incoherent beliefs.

The acts of supposing A to be the case and negatively supposing A are also incoherent with one another. For strategic reasons, it
can sometimes make sense to make incoherent suppositions, as, for
example, when we want to show that some statement is false. But
when we begin by making incoherent suppositions, and subsequently
eliminate them, we are "honoring" the general commitment to achieve
coherence.

If we suppose A (we perform the act represented: $\llcorner A$) and reason
to incoherence, then if all other premises are assertions or denials,
we are entitled to discharge the supposition and conclude with either
of the following illocutionary acts:

$$\dashv A \qquad \text{or:} \qquad \vdash \sim A$$

(If one of the remaining premises is a supposition, then the conclusion
will be either illocutionary act shown here:

$$\neg A \qquad \text{or:} \qquad \vdash \sim A \quad)$$

However, if we begin by negatively supposing A, and reason to incoherence, there is some dispute about what moves we are entitled to make. If negatively supposing A leads to incoherence, then positively supposing the negation of A ($\llcorner\sim A$) will also lead to incoherence. So if negatively supposing A and positively supposing $\sim A$ both lead to incoherence, we are clearly entitled to discharge the supposition, and conclude with either of the following results:

$$\vdash \sim\sim A \qquad \text{or:} \qquad \dashv \sim A$$

But may we also legitimately conclude by asserting A? Before considering what is the right answer to this question, we should note that if the right answer is "Yes," we will get an illocutionary system of classical logic. If it is "No," we get an illocutionary system of Intuitionist logic. The different roles that negation plays in the two systems are consequences of the different roles that negative supposition plays in these systems.

To properly determine what we can conclude when a negative supposition leads to incoherence, I think we must distinguish *factual* from *fictional* statements. A factual statement is one that is designed to fit the actual world. It is true if it does fit, and false if it fails to fit the world. A speaker who produces a factual statement may not intend for her statement to fit the world, but she must intend for her statement to be "measured" against the actual world.

A fictional statement is not a statement made by a character in a story, it is a statement made *about* characters and events in a story, or "story world." Real people make both fictional and factual statements. Whether a statement is factual or fictional depends on a language user's intentions. Someone telling a story intends to make fictional statements. If his listeners or readers understand the speaker/writer to be telling a story, then in talking about the characters and events in the story (in the "story world"), they will intend their statements to be fictional. However, a person can, by mistake, regard statements from a story to be concerned with the actual world, with actual people and actual events. When this person makes statements about the characters and events in the story, she intends to be talking about actual people and events. She is mistaken, but she intends for her statements to be measured against the actual world.

She is either making factual statements that don't "measure up," or trying but failing to produce factual statements. She isn't making fictional statements if she doesn't think she is.

Not only must we recognize both factual and fictional statements, but we must also recognize different modes of assertion and denial. When we make a *factual assertion*, we accept a factual statement as being, or representing, what is the case. Different fictional statements concern different stories, or story worlds. To *fictionally assert* a fictional statement is to accept that statement as an authorized or authoritative representation of the way things are in the story world.

There isn't really a story world, or worlds, against which fictional statements can be measured. But fictional statements can be true or false. To discuss this issue further, I will confine my attention to coherent fiction: this is fiction which doesn't endorse inconsistent statements—it doesn't commit its readers to both assert and deny some one fictional statement. Whether or not a fictional statement is true is determined by what the story says, by some default assumptions, and by what follows from the story and the default assumptions.

Certain default assumptions are determined by genre conventions. The conventions for science fiction stories are different from those for romance novels, for example. Horror stories and vampire stories have their own conventions. Within limits determined by genre conventions, one important default assumption is that, unless the story says otherwise, the story world is like the actual world in most ordinary respects. What is true of a given story world is what an ideal reader of the story or stories, a reader who knows the whole story or all the stories, and is also aware of the default assumptions, is committed to fictionally assert.

Not all fictional statements are either true or false. This is because, among other reasons, it isn't possible for any human author to tell a complete story. If we consider Sherlock Holmes' mother's mother's mother, and ask whether this great grandmother had blue eyes, there is no answer to the question. Nothing in the stories or default assumptions determines that her eyes were blue, and nothing determines that they weren't. It isn't true that she had blue eyes, and it isn't false either. But they weren't both blue and not blue, they couldn't be.

If we are dealing with a language for which some statements fail to be either true or false, and we understand a disjunctive statement to be true iff at least one disjunct is true, then some instances of excluded middle will fail to be true or false. We can't correctly assert/accept such statements. We can't deny them either. For we can't rule out their assertion for objective reasons. If we consider fictional statements, a statement that fails to be true or false might become one or the other if the story were continued. The story doesn't rule out our accepting the statement, it just doesn't say enough to decide the matter.

But even if A is a statement that is neither true nor false, a statement '$[A \& \sim A]$' is one we are entitled to deny. For we can rule out the assertion of this statement. No coherent further development of the story will make this a true statement. Since we can correctly deny a contradictory statement, then, by our principle $\sim$ *Introduction*, we can correctly assert that the statement is false.

If we suppose a *factual* statement to be false, and then end in incoherence, I think we are entitled to take back our supposition and accept the factual statement. The world is determinate and complete, which is not to say that what is present and past completely determines what will happen in the future. Factual statements either fit the world or they don't. If positively supposing a fictional statement leads us to incoherent consequences, then that statement can't be true; we are surely entitled to deny, to rule out the assertion of, the fictional statement; this means that the statement is false of the story world.

However, if negatively supposing a fictional statement leads to incoherence, then we can't be confident that the statement is true of the story world. In such a case we can deny the statement's negation, and can also assert that the statement's negation is false. But we don't accept this principle:

$$\frac{\dashv/\neg \sim A}{\vdash/\llcorner A}$$

or this one:

$$\frac{\vdash/\llcorner \sim \sim A}{\vdash/\llcorner A}$$

I concede that an illocutionary system of intuitionist logic is correct for assertions and denials of statements of coherent fiction, while holding out for a classical system of logic for factual assertions and denials. If mathematics is best understood to be a highly sophisticated form of fiction, then perhaps it is a system of intuitionist logic which represents the logic of mathematics.

I might be wrong that factual statements all either fit or fail to fit the world. But I am not wrong in my understanding of assertion and denial. These are fundamental and inescapable speech acts, no matter what language a person speaks. The characterization of negation in terms of denial leaves some room for developing alternative concepts, or versions, of negation. But it doesn't leave room for a version of negation such that both a statement and its negation can legitimately be asserted. For then it would be legitimate to both assert and to deny a single statement. But as we have understood assertion and denial, this can make no sense. We might do one of these, then change our minds, and do the other, but we would be "taking back" the act we first performed.

We might also assert a statement on one occasion, and deny it on another, without realizing that we had ended up in incoherence. But this is a sign of error. When the incoherence is brought to our attention, we recognize that we need to give up on either the assertion or the denial. We can't, on a single occasion, both do something and refrain from doing it. And we can't successfully direct a person to both do something and refrain from doing it. Neither can we, on a single occasion, both accept a statement and rule out its acceptance. There is no way to make sense of *paracoherence* with respect to illocutionary acts.

John T Kearns
Department of Philosophy and Center for Cognitive Science
University at Buffalo, SUNY
Buffalo, New York 14260
e-mail: `kearns@buffalo.edu`

A Notion of Concept is either Superfluous or Procedural

Pavel Materna*

Abstract

It is shown that a set-theoretical (in particular, Fregean) definition of concepts implies that whatever is claimed using the term 'concept' can be claimed using the term 'class' and that the only remedy consists in construing concepts as a kind of abstract procedure. This has been already done within Transparent intensional logic. The author argues that this approach to concepts would have satisfied Kurt Gödel, who demanded setting up "a consistent theory of classes and concepts as objectively existing entities".

1 An explication is needed

The term *concept* is frequently used in colloquial English a well as in various professional texts. No doubt the meaning of this term varies and the frequency of its use causes that it lacks any definite meaning. In such a situation an *explication* is needed.

There are two great areas where the term is used: *cognitive sciences* (in particular psychology) and *philosophical logic* (including *logical analysis of natural language, LANL*). For the former concepts are *mental entities*.[1] Here we set aside the possible explications. For philosophical logic *concepts are objective non-mental and*

---

*This contribution has been supported by Grant Agency of Czech Republic, project No P401/10/0792, "Temporal aspects of knowledge and information".

[1] Jerry A. Fodor: *Concepts are mental particulars; specifically, they satisfy whatever ontological conditions have to be met by things that function as mental causes and effects* (Fodor, 1998, p. 23)

The (natural) numbers x greater than 1 divisible just by 1 and x.
The (natural) numbers that possess just two factors.

Sinn$_1$

Sinn$_2$

Denotation (*Bedeutung*)
(i.e., the set of primes)

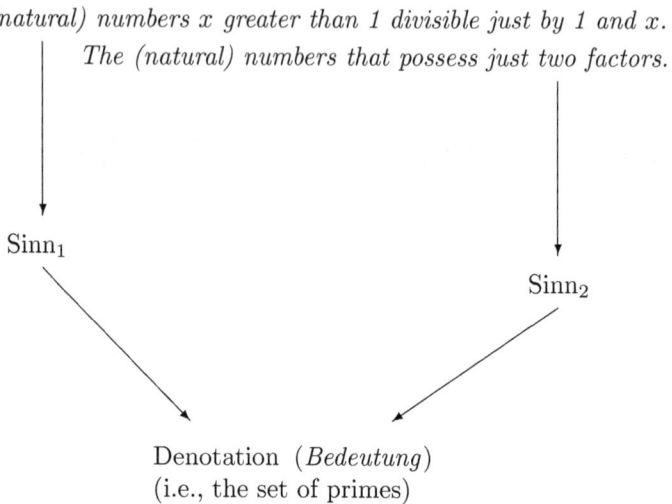

Figure 1: Two (equivalent) definitions of prime numbers

extra-linguistic entities that can be associated with expressions as their meanings. Our explication concerns this basic view.

According to Carnapian explication we have to find out some basic features that are shared by such uses of the term that we are disposed to accept as relevant. Now we will adduce some basic intuitions that should help to start our explication.

1.1 Concepts can be associated with expressions as their meanings

Alonzo Church articulated this intuition in his (Church, 1956) when correcting Frege's definition of concept. Church proposed that **the** *sense of an expression E is **a** concept of the denotation of E*, where *sense* is what Frege called *Sinn* and what we can (and mostly will) call *meaning*.

As an example consider two (equivalent) definitions of prime numbers in Figure 1. Observe that we have for each expression just one meaning (therefore *the* sense) but more than one concept of the denotation (therefore *a* concept). Note also that concepts are for Church on the *Sinn*-level of the famous Frege's semantic schema.

1.2 Concepts are not expressions

Accepting the preceding intuition we automatically distinguish expressions and concepts. Independently thereof we can see that this intuition is very natural. Which kind of expression could be detached as being a concept? In all well-known theories of concept the assumption that there are expressions of a language and that these expressions can be associated with concepts is self-evident, even implicit. After all, a most natural definition of synonymy has it that synonymous expressions share a concept.

1.3 Not every concept is universal

the even number, the highest mountain, the Pope, ... are expressions that do not express universal concepts but they are surely supposed to express *some* concept. Also such expressions as *quickly, some, because, if, ...* deserve to be connected with some concept.

(Bolzano's theory of concepts in his Bolzano, 1837 is in full harmony with this intuition.)

1.4 To understand an expression = to know the respective concept

To hold this intuition we try to explain the central phenomenon studied by semantics: how come that we *understand* expressions of the given language?

Expressions *denote* objects.[2] They do it mostly not immediately, as labels, but due to indicating how to get the denotation. They not only *denote*, they *mean* something. We understand them in virtue of this *meaning*, one explication of which is just *concept*. This distinction *denotation* vs. *meaning* makes up the germ of Frege's semantic 'triangle'. So the next intuition can be formulated.

1.5 A concept is a 'way' to an object, in particular to the denotation of the given expression

This is a good metaphor. Surely, there is no intrinsic relation between the sequence of sounds/letters and the object denoted. A concept can

---

[2]By *objects* we do not mean particular perceivable objects like particular trees etc. For example *tree* denotes a property.

be seen just as such a 'bridge' leading from the expression *qua* such a sequence to the object denoted. The intuition can continue as follows: the particular bricks that such a bridge is composed of should somehow correspond to particular components of the respective expression.

1.6   In principle, concepts are structured

This intuition is closely connected with the preceding one. In principle, *expressions* are structured. If concepts should connect expressions with objects and be at the same time simple, unstructured, their connecting role would be fully enigmatic.[3]

Now we will show that the well-known Frege's definition of concept is incompatible with most of the basic intuitions connected with the term *concept*, further that it makes the notion of concept superfluous because replaceable by the notion of class, and finally that attempts to correct Frege's notion of concept and retain the set-theoretical approach cannot succeed.

2   A set-theoretical approach. Frege's definition

The historically first theory of concepts is probably *Aristotle's theory of definitions*.[4] A set-theoretical interpretation (frequently used in traditional textbooks) is thinkable but not necessary. It can be shown that it is more friendly to our intuitions than Frege's theory. (An interesting justification of this claim can be found in Tichý, 1968, p. 81 in Tichý, 2004, where *meaning* or *concept* should replace Tichý's term *intension*.)

Another logician, whose theory of concepts is not unambiguously set-theoretical is Bolzano in (Bolzano, 1837), whose *Begriffe* can be construed as being structured.

Yet the first logician whose theory of concepts has been appreciated by modern logic is Gottlob Frege. In his (Frege, 1891, 1892) and elsewhere he has placed his concepts in his schema on the level of denotation (*Bedeutung*) and defined them as (total) characteristic functions of classes (as we would say it nowadays).

---

[3]This is a great problem even for cognitivists: see Fodor's attempts at defending his conceptual atomism in his (Fodor, 1998).

[4]More details (from our viewpoint) can be found in (Materna & Petrželka, 2008).

Now we will articulate some essential objections to Frege's definition and argue that such objections can be raised to any set-theoretical definition.

2.1 In any context where Frege uses and would use the term 'concept' the term 'class' could be used salva veritate.

Consider, e.g., the famous pages of *Grundlagen der Arithmetik* (Frege, 1884), where numbers are associated with 'concepts', where Frege means by concepts just what he seven years later defined as (properly speaking) classes. (Read 'class' wherever you see 'Begriff'.)

So *why should we use two terms*? Says Ockham's Razor.

2.2 Any attempt to conceptually capture non-universals is ruled out ex definitione.

Among our basic intuitions there was *Not every concept is universal*. While the concept underlying the expression *mountain* is a universal concept and can be therefore classified as such by Frege the concept expressed by *the highest mountain* is not a universal concept and Frege has to identify such concepts with singletons. Saying however that the highest mountain is in Asia we do not mean that the singleton containing the highest mountain is in Asia.

Not only that: why should expressions that do not denote universals do without concepts? The adverbials, for example, are surely meaningful. Frege however would not connect such expressions (e.g., *quickly*) with concepts.

Another (not too essential) point: Frege does not classify *relations* (*Beziehungen*) with concepts (see, e.g., Frege, 1971, pp. 150–151). This can be easily repaired, though.

2.3 Fregean concepts cannot be 'ways' to an object.

There are two reasons of this shortcoming:

a) Frege's concepts are on the level of *Bedeutung*.

b) They are (characteristic functions of) *classes*.

Ad a): It was Church who recognized in (Church, 1956) that this position of concepts is highly unnatural, perhaps counterintuitive. For

Church any good explication of *concept* presupposes that concepts are not primarily what is denoted: they should correspond to Frege's *Sinn* rather than to his *Bedeutung*. (Church's theory admitted however that concepts can be sometimes on the level of *Bedeutung*, viz. when we speak about them, i.e., mention them.) The proper role of concepts is that they *determine* some object.

Ad b): No set-theoretical object can be construed as being something like 'way to'. Remember Zalta in (Zalta, 1988): *There is nothing about a set in virtue of which it may be said to present something to us.*

The next objection is a vivid specification of the present one.

2.4 Concepts of mathematical classes are the classes themselves.

Returning to our recent example let us consider two expressions:

i) *The numbers x greater than 1 divisible just by 1 and x*

ii) *The numbers that have just two factors*

Both i) and ii) denote the class of primes. According to Frege's definition there is one concept here, viz. this very class. According to our intuitions (and to Church), i) *expresses* a concept C, ii) *expresses* a concept C', and $C \neq C'$. Thus Frege could not explain what we do: the semantic distinction between i) and ii) consists in expressing two distinct concepts. Thus he lets pass the opportunity to make concepts responsible for such cases of semantically equivalent but distinct expressions. The role of his concepts is therefore enigmatic.

2.5 Frege's concepts are not structured.

Set-theoretical objects like (characteristic functions of) classes are simple. If the semantics of the preceding expressions for primes reduces to the set of primes then it is impossible to derive it from the structure of the given expression. Frege's concepts are then semantically irrelevant and can serve at most as a double of classes.[5]

---

[5]Well, one might object that Frege's functions could be interpreted as 'functions-in-intension', i.e., as procedures creating functions like our *constructions* (see ch. 3) while *Wertverlauf* would be a set, but no definite details, which would support a different view of the point 2.5, are available.

2.6   A mentioned concept is no more a concept.

Frege's concepts are functions and are, therefore, 'unsaturated'. Thus if concepts are talked about (mentioned) they cannot be denoted by 'concept words' in the position of a grammatical subject and they become objects (*Gegenstände*). (See Frege, 1892, discussion with Kerry.) Frege cannot say that a concept remains to be a concept even if it is mentioned. This is not very intuitive.

2.7   No special treatment of empirical concepts is provided for.

This is not an objection concerning just Frege. First, no adequate means for semantically distinguishing empirical and non-empirical concepts was at Frege's disposal then. Second, the same objection can be reproached to many contemporary semanticists. In (Duží, Jespersen, & Materna, 2010, p. 303) we can read "The problem with neglecting the difference between empirical and non-empirical is that the specifically *empirical modality of contingency* is swept under the carpet."

Summing up:

At least the drawbacks given by objections 2.1, 2.3, 2.4, 2.5 cannot be repaired using a set-theoretical approach.

    A key objection is 2.1: Why do we need two terms: *class* and *concept*? If only for distinguishing the case where the concept is mentioned (objection 2.6) then this would be a very weak motivation.

    Any set-theoretical attempt to answer the objections 2.3–2.5 must break down. Ad 2.3: no set-theoretical object can play the role of a 'way' to something. Ad 2.4: Which set-theoretically defined object could identify a class K of mathematical objects and not be identical with K? Ad 2.5: *Ex definitione.*

    Thus the main drawback connected with Frege's theory of concepts consists in the fact that it is based on a coarse-grained (=set-theoretical) approach.

    Observe that nothing would change in this respect if we tried to answer the objection 2.7 by introducing PWS intensions: these are also sets, being functions (mappings) from possible worlds. If concepts should play the role of 'ways' to an object then they could not be intensions. Besides, intensions cannot be used when mathematical objects have to be semantically analyzed.

3   A procedural approach

We have said that PWS intensions are still set-theoretical objects. So intensions are too coarse-grained. Thus what we need can be called *hyperintensions*, and this term is used in Cresswell's (Cresswell, 1975). Yet the history of relevant attempts at reaching a hyperintensional solution of some semantic problems is much older.

Carnap in his (Carnap, 1947) recognized that his 'method of intensions and extensions' broke down in certain kind of contexts ('indirect contexts', in particular attitudinal contexts). The well-known history continues as follows: Carnap himself, looking for such fine-grained relations between expressions that would help to explain why analytically/logically equivalent expressions can all the same semantically differ, proposed a definition of *intensional isomorphism*, which in some way respected structure of an expression. **Church** succeeded in showing that Carnap's definition is not satisfactory and tried to define (in fifties and later) such a fine-grained criterion which would not share drawbacks of Carnap's attempt. Church's final version (Church, 1993) of the alternatives of 'synonymous isomorphism' formulates Alternatives: Alternative 0, 1, 2 and 1', and inclines to Alternative 1 or 1', both are based on λ-convertibility, Alternative 1 on α-rule and β-abstraction and β-reduction, Alternative 1' adds η-reduction. We can state here that the relation of *procedural isomorphism*, which enables us to procedurally define concepts, corresponds essentially to Alternative 1' minus β-conversion. For more details see (Jespersen, 2010).

Cresswell's tuples:

In his (Cresswell, 1975) M. J. Cresswell proposes the following solution to the problem of finding such a definition of meaning which would make meaning mirror the structure of the respective expression: Let δ be a functor (in the given language) and $\alpha_1, \ldots, \alpha_n$ arguments. Let α be an expression possessing the structure $\langle \delta, \alpha_1, \ldots, \alpha_n \rangle$. Then

$$M(\text{eaning of})(\alpha) = \langle M(\delta), M(\alpha_1), \ldots, M(\alpha_n) \rangle$$

i.e., "The meaning ... is ... the $n + 1$-tuple consisting of the meaning of the functor together with the meanings of its arguments". Thus one goal was achieved: the components of such tuples mirror the components of the given expressions. Yet no structured meaning has been defined; only particular components of the meaning have got a definition.

Interestingly, in (Bolzano, 1837) Bolzano has warned against confusing concepts with their contents. A concept is according to him the *way* (*die Art, wie*) in which the components of its content (*Merkmale*) are interconnected (Bolzano, 1837, I., p. 244). This *way* is what is missing in Cresswell's tuples. (See Tichý, 2004, pp. 835–841, or Jespersen, 2003.)

Bealer (1982):

George Bealer in his inspiring (Bealer, 1982) has shown that (p. 2)

> There have been two fundamentally different conceptions of properties, relations, and propositions. On the first conception intensional entities are considered to be identical if and only if they are *necessarily equivalent*... On the second conception... each definable intensional entity is such that when it is defined completely, it has a *unique, non-circular definition*.

Therefore he divided 'intensions' into two classes: for *qualities, connections* and *conditions* it holds that they are identified *via* necessary equivalence (in the parlance of TIL these are *properties, relations* and *propositions* respectively) while necessary equivalence is not sufficient to identify *thoughts* and *concepts*. Bealer has recognized that we have to distinguish what TIL calls "functional (intensional) level" and "conceptual (hyperintensional) level".[6]

Tichý (1968, 1988):

Pavel Tichý (1968) (Filosofický časopis) "Smysl a procedura" (Sense and Procedure, Tichý, 2004) called our attention to the fact that

---

[6]It is another question in which way Bealer logically handled the distinction.

while the notion of effective *procedures* was successfully used in application to syntactic problems its applicability to *semantics* was underestimated. Actually, however, "the relation between sentences and procedures is of a semantic nature; for sentences are used to record the results of performing particular procedures" (Tichý, 2004, p. 80). Tichý illustrated this claim by showing (and in 1969 in *Studia Logica* detailing) that fundamental semantic notions can be explicated in terms of Turing machines.

This can be considered as the beginning of a *genuine procedural semantics*. The development of the incipient idea of *Transparent intensional logic* ("TIL") by Tichý culminated in (Tichý, 1988) (*The Foundations of Frege's Logic*), where the fundamental notions of TIL have been introduced. The present state of TIL is described in (Duží et al., 2010).

TIL in brief:

The key step towards making semantics procedural (and thus structured) consists in defining *constructions* as it has been realized in TIL.

Informally, TIL constructions are *abstract procedures*. They are abstract, since not localizable in space and time. (The *tokens of records of constructions* are, of course, concrete entities, in principle always localizable.) They are also *extra-linguistic* and can be—as a result of linguistic convention—associated with some sequences of characters/sounds as their *meanings*.

The most frequently used collection of constructions in TIL is (omitting precise definitions that can be found in Tichý, 1988 or, e.g., in Duží et al., 2010):

Variables:

The objects that the expressions of a given language can talk about are type-theoretically classified. (See below.) For each type there are countably infinite many variables at our disposal. As a kind of constructions they are extra-linguistic and the letters we are using to denote them ($x_i, y_i, \ldots, p_i, q_i, \ldots, m_i, n_i, \ldots$) are just names of variables. Variables, as well as constructions containing (free) variables, construct objects dependently on *valuations*, so they v-construct objects, where v is a parameter of valuations.

Trivializations:

Let X be an object (or even a construction). ^{0}X is a *Trivialization*. It *mentions* X (not *using* it) and so it leaves X without any change.[7]

Compositions:

Composition (v-)constructs the value of a function constructed[8] by X on arguments constructed by X_1, ..., X_m. Record: $[XX_1 \ldots X_m]$.

Closures:

Closure constructs a function by abstraction. Record: $\lambda x_1 \ldots x_m X$, where x_i are pairwise distinct variables v-constructing objects of types β_1, ..., β_m, respectively, and X v-constructs objects of the type α. The closure constructs a function from $\langle \beta_1, \ldots, \beta_m \rangle$ to α in the manner well-known from λ-calculi. (As for a precise definition, see again Duží et al., 2010.)

Some further constructions may be defined, in TIL a useful construction *Double Execution*, ^{2}X, realizes a construction twice.

Formally, the inspiration is clear: a typed λ-calculus, which is a most suitable tool for treating functions. Church's ingenious idea that operations on functions are essentially reducible to *creating functions* and *applying functions to arguments* has been exploited in TIL and, independently, in Montague's logic (See Montague, 1974).

Now we have to define types in terms of which we classify the area of objects the given language can talk about. First, types of order 1 are defined:

---

[7]The impression as if such a construction were useless is misleading: Trivialization is extremely important, it makes it possible to ascend from intensional to hyperintensional level.

[8]We will write 'construct' admitting however that actually v-constructing takes place.

Types:

Types of order 1: Let B be a *base*, i.e., a non-empty collection of disjoint non-empty classes.

  i) The members of B are *types of order 1 (over B)*[9]

  ii) Let α, β_1, ..., β_m be *types of order 1*. Then $(\alpha\beta_1 \ldots \beta_m)$, i.e., the class of partial functions with values in α and arguments in β_1, ..., β_m, respectively, is a *type of order 1*.

  iii) Nothing is a *type of order 1* unless it obeys i), ii).

Second, *higher-order types* must be defined, otherwise hyperintensionality cannot be realized.

Higher-order types: Ramified hierarchy of types
In brief: Given types of order 1 we define *constructions of order n*. The main idea is that a construction of order n constructs an object whose type is of order n. Then: let $*_n$ be the set of constructions of order $n : *_n$ and types of order n are *types of order n + 1*. Finally, if α, β_1, ..., β_m are *types of order n + 1* then $(\alpha\beta_1 \ldots \beta_m)$ is a type of order $n + 1$.

The definition of higher-order types is of key importance: it makes it possible to not only *use* but also *mention* constructions. Now we do not need to use meta-language when wanting to talk about constructions. The resulting logic is *hyperintensional*.

The most frequently used *base* in TIL determines which objects a natural language can be talked about. It contains following simple types:

o ...... $\{\mathbf{T}, \mathbf{F}\}$, i.e., the set of truth-values $\mathbf{T}$ (truth), $\mathbf{F}$ (false)

ι ...... the set of (bare) individuals

τ ...... the set of real numbers, doubling the set of time moments

ω ...... the logical space, i.e., the set of *possible worlds*.

---

[9]We will omit "over B" henceforth.

Example 1   The construction that constructs **T** and can be ascribed to the sentence *2 is a prime* as its meaning:
Types:[10] $2/\tau, \mathrm{Prime}/(o\tau)$

$$[{}^0\mathrm{Prime}\ {}^02]$$

The construction that constructs a proposition (type $((o\tau)\omega)$, abbr. $o_{\tau\omega}$):

Charles knows that 2 is a prime

Types: as above + Charles/ι, know/$(((o\iota*_1)\tau)\omega)$, abbr. $(o\iota*_1)_{\tau\omega}$, $w \to \omega, t \to \tau$.
Instead of $[[Xw]t]$ we write X_{wt}.

$$\lambda w \lambda t [{}^0\mathrm{know}_{wt}{}^0\mathrm{Charles}\ {}^0[{}^0\mathrm{Prime}\ {}^02]]$$

In this second example the construction $[{}^0\mathrm{Prime}\ {}^02]$ is *mentioned*: While in the first example this construction is *used* to construct **T**, in the second example the construction is not realized: to construct the proposition it does it without taking into account which truth-value is constructed by $[{}^0\mathrm{Prime}\ {}^02]$. Therefore we have to trivialize this construction, which blocks any *use* of it.

Concepts[11]

For TIL, the *meaning* of an expression is a construction. To identify the meaning of an expression with a procedure (a TIL construction) means to be in harmony with basic intuitions and to repair the drawbacks mentioned in the above list of objections to Fregean concepts-as-classes.

Remark 1   Our previous criticism of Frege's definition of concepts is based on the following assumption: Frege's characteristic *function* of a class has to be interpreted as a function *qua* mapping. Only some inconspicuous formulations in Frege seem to indicate that it should be otherwise. Yet some places in (Tichý, 1988) suggest that we could sometimes interpret Frege's *function* as a way how to obtain

---

[10]We write X/α to say that X is (a member) of the type α.

Where C is a construction we write $C \to \alpha$ to say that the object constructed by C is of the type α.

[11]See (Materna, 1998, 2004).

such a mapping, and explain unclear points in Frege's philosophy as consequences of his oscillating between set-theoretical and procedural viewpoint. If Frege's *function* were construed as a procedure (and, therefore, *Wertverlauf* as the respective mapping) then much of our criticism would have to be revised, of course.

First of all we have to distinguish two cases:

a) *its color*

b) *a speckled dog*

We would like to associate the b) case with a *concept*: it will be a construction that constructs a property, viz. the property of being a speckled dog. As for a), we get a procedure which will expect a completion to determine the individual whose color is mentioned. This case represents the expressions that contain some *indexicals*, which will be represented by free variables in the respective construction. The case b) represents expressions without indexicals: the expressed constructions will be closed, i.e., will not contain free variables. It is this latter kind of expressions that can be said to express *concepts*:

Concepts are closed constructions.

This is however just an approximation. For consider the expression

(E) *number(s) greater than 0*

There are two so-called *simple* concepts here (where a simple concept C contains just C as its constituent, i.e., a used subconstruction): $^0>$, ^{00}, variables ranging over real numbers).

Following constructions are candidates for being the concept expressed by (E):

$$\lambda x_1[^0> x_1{}^00], \lambda x_2[^0> x_2{}^00], \ldots, \lambda x_k[^0> x_k{}^00], \ldots$$

There are infinitely many members of this sequence. Does it mean that the expression (E) expresses infinitely many concepts? And which English expressions would correspond to particular variants?

We need some kind of 'normalization', which would select one representative of the sequence whose members are pairwise α-*equivalent*, i.e., differ just by λ-bound variables.[12]

---

[12] In TIL we distinguish two kinds of binding variables. The second one consists in occurrence of the variable in a mentioned construction, i.e., in ^0C, where C is a construction.

Further, one representative of a sequence of pairwise *η-equivalent* constructions can be determined, where a construction C is *η-equivalent* to a construction D iff C is an *η-reduction* of D (so D is an *η-expansion* of C). An example (Believe/$(o\iota o_{\tau\omega})_{\tau\omega}, x \to \iota, y \to o_{\tau\omega}, t \to \tau, w \to \omega$):

$$^0\text{Believe}; \lambda w[^0\text{Believe } w]; \lambda w \lambda t \, ^0\text{Believe}_{wt};$$
$$\lambda w \lambda t[\lambda xy[^0\text{Believe}_{wt}xy]], \dots$$

As a result we can define *procedural isomorphism* as follows:

A construction C is *procedurally isomorphic* to a construction D iff there exists a sequence $C_0, \dots, C_m$ such that C_0 is C, C_m is D, and for any C_j, C_{j+1} it holds that C_j is α- or η-equivalent to C_{j+1}.

The process of normalization that selects one definite concept from among procedurally isomorphic constructions is described in (Horák, 2002). The final definition is:

Concept is a closed construction in normal form.

In our examples $\lambda x_1[^0 > x_1{}^0 0]$ and ^{0}Believe are concepts, the other constructions from the mentioned sequences are said to 'point to' the concept.

As soon as concepts are defined as a kind of constructions they are no more set-theoretical objects and we can make sure that they satisfy our basic intuitions, and that our objections to Frege's definition cannot be applied to our procedural definition.

Appendix

Gödel's problem with concepts

In his article *Russell's Mathematical Logic* (Gödel, 1944, reprinted in Gödel, 1990, pp. 119–141) Gödel says:

> I shall use the term "concept" in the sequel exclusively in this objective sense. One formal difference between the two conceptions of notions would be that any two different definitions of the form $\alpha(x) \equiv \phi(x)$ can be assumed to define two different notions α in the constructivistic sense.

...For concepts, on the contrary, this is by no means
the case, since the same thing may be described in different
ways. ...There is certainly more than one notion in the
constructivistic sense satisfying this condition, but there
might be one common "form" or "nature" of all pairs.
(see Gödel, 1990, p. 128)

The way Gödel used the notion concept here is not at all clear.[13]
This was noticed also by Ch. Parsons in his comment (Parsons, 1990,
p. 110):

By 'concepts' Gödel evidently means objects signified
in some way (italics mine—Materna) by predicates. ...
Gödel's remarks about realistic theories of concepts ...
have an inconclusive character, no available theory satisfies
him.

Let us return to our example with prime numbers. Consider the
following equalities:

$$Prime(x) = \text{ The (natural) numbers } x \text{ greater than 1 divisible just}$$
$$\text{by 1 and } x \tag{1}$$
$$Prime(x) = \text{ The (natural) numbers that possess just two factors} \tag{2}$$

For 'constructivists'[14] there are two concepts here (represented by
the two right sides of these equalities. For Gödel there is just one
concept here (*somehow* represented by the common left side).

Clearly, Gödel refused to accept the 'constructivist' view but was
not ready to propose such a definition of concepts which would satisfy
his realistic philosophy. His refusal of constructivists view was based
on realism: the constructivists were for him nominalistic subjectivists.
His recommendation "*to make the meaning of the terms 'class' and
'concept' clearer and to set up a consistent theory of classes and concepts as objectively existing entities*" can be considered as being a sort
of legacy.

If our claim that his refusal of constructivists was motivated by
realistic philosophy is right then we are entitled to claim that if Gödel

---

[13] Portions of this section draw on material presented in (Materna, 2007).
[14] But also, e.g., for (Bolzano, 1837; Bealer, 1982).

had accepted (with TIL) that constructions are objective entities then he would have revised his negative standpoint to constructivists' view.

Going still further I dare to claim that *the procedural conception of (not only) concepts in TIL is suitable to fulfill the expectations expressed by the above Gödel's recommendation.*

References

Bealer, G. (1982). *Quality and Concept.* Oxford: Clarendon Press.

Bolzano, B. (1837). *Wissenschaftslehre.* Sulzbach.

Carnap, R. (1947). *Meaning and Necessity.* Chicago: Chicago University Press.

Church, A. (1956). *Introduction to Mathematical Logic.* Princeton: Princeton University Press.

Church, A. (1993). A Revised Formulation of the Logic of Sense and Denotation. Alternative (1). *Noûs, 27*, 141–157.

Cresswell, M. J. (1975). Hyperintensional logic. *Studia Logica, 34*, 25–38.

Duží, M., Jespersen, B., & Materna, P. (2010). *Procedural Semantics for Hyperintensional Logic. Foundations and Applications of Transparent Intensional Logic.* Springer.

Fodor, J. A. (1998). *Concepts.* Oxford: Clarendon Press.

Frege, G. (1884). *Die Grundlagen der Arithmetik.* Breslau: W. Koebner.

Frege, G. (1891). *Funktion und Begriff.* Jena: H. Pohle.

Frege, G. (1892). Über Begriff und Gegenstand. *Vierteljahrsschrift für wissenschaftliche Philosophie, 16*, 192–205.

Frege, G. (1971). *Schriften zur logik und sprachphilosophie. aus dem nachlass* (G. Gabriel, Ed.). Hamburg: Felix Meiner Verlag.

Gödel, K. (1944). Russell's mathematical logic. (In Gödel, 1990, pp. 119–141)

Gödel, K. (1990). *Collected Works* (Vol. 2; S. Feferman, J. W. D. Jr., S. C. Kleene, G. H. Moore, R. M. Solovay, & J. van Heijenoort, Eds.). Oxford: Oxford University Press.

Horák, A. (2002). *The Normal Translation Algorithm in Transparent Intensional Logic for Czech.* Unpublished doctoral dissertation, Masaryk University, Brno. Available from `http://www.fi.muni.cz/~hales/disert`

Jespersen, B. (2003). Why a tuple theory of structured propositions isn't a theory of structured propositions. *Philosophia, 31*, 171–183.

Jespersen, B. (2010). Hyperintensions and procedural isomorphism: Alternative (1/2). In K. Kijania-Placek (Ed.), *Proceedings of ECAP VI.* London: College Publications.

Materna, P. (1998). *Concepts and Objects* (Vol. 63). Helsinki: Philosophical Society of Finland.

Materna, P. (2004). *Conceptual Systems.* Berlin: Logos Verlag.

Materna, P. (2007). Properties of mathematical objects (Gödel on classes, properties and Concepts). *Journal of Physics: Conference Series, 82*, 1–15.

Materna, P., & Petrželka, J. (2008). Definition and concept. Aristotelian definition vindicated. *Studia Neoaristotelica, 5*(1), 3–37.

Montague, R. (1974). *Formal philosophy: Selected papers of richard montague* (R. Thomason, Ed.). New Haven: Yale University Press.

Parsons, C. (1990). Introductory note to 1944.
(In Gödel, 1990, pp. 102–118)

Tichý, P. (1968). Smysl a procedura. *Filosofický časopis, 16*, 222–232. (Translated as 'Sense and procedure' in Tichý, 1968, pp. 77–92)

Tichý, P. (1969). Intensions in terms of Turing machines. *Studia Logica, 26*, 7–25. (Reprinted in Tichý, 2004, pp. 77–92)

Tichý, P. (1988). *The Foundations of Frege's Logic.* Berlin, New York: De Gruyter.

Tichý, P. (2004). *Collected Papers in Logic and Philosophy* (V. Svoboda, B. Jespersen, & C. Cheyne, Eds.). Prague and Dunedin: Filosofia and University of Otago Press.

Zalta, E. (1988). *Intensional Logic and the Metaphysics of Intentionality.* Cambridge, London: MIT Press.

Pavel Materna
Institute of Philosophy, Academy of Sciences of the Czech Republic
Jilská 1, 110 00 Prague 1, Czech Republic
e-mail: materna@lorien.site.cas.cz

The Display Problem Revisited

Tyke Nunez*

Abstract

In this essay I give a complete join semi-lattice of possible display-equivalence schemes for display logic, using the standard connectives, and leaving fixed only the schemes governing the star. In addition to proving the completeness of this list, I offer a discussion of the basic properties of these schemes.

1 Introduction

In this essay I will build on the work begun in (Belnap, 1996). In his essay Belnap presented various options for how to set up the display equivalences of display logic, a refinement of Gentzen's sequent calculus.[1] Belnap describes most of the schemes I will deal with, although I will complete the list he presents, using his connectives, and leaving fixed only the schemes governing the star, as he does.[2]

Even so, I hope that having a complete lattice of display equivalence schemes will allow a more systematic understanding of the properties these give to the structural-connectives, whose meaning seems not to be well understood. These properties will be similar to the properties introduced by structural rules, the other kind of rule

---

*Thanks to audiences at Logica 2010 and the University of Pittsburgh, especially Nuel Belnap, Shawn Standefer, Kohei Kishida and Kathryn Lindeman, for comments on the presentation associated with this essay as well as this essay itself.

[1](Gentzen, 1969). Regrettably, I am unaware of an 'easy introduction' to display logic. Belnap's original paper (Belnap, 1982), and its subsequent slightly updated version in §62 of (Anderson, Jr, & Dunn, 1992) are, I think, the best places to start.

[2]Where it isn't a detriment to clarity I will take up Belnap's names for the schemes and skip over points he has already made.

in display logic governing the structural-connectives. The chief difference between these rules and the equivalence schemes is that the latter must secure the display property, which is, roughly, that any structure can be 'displayed' alone as either the entire antecedent or consequent of a consecution display equivalent to the original one.[3] This property, the defining feature of display logic, sets a restriction on the possible schemes of display logic; securing this property is what Belnap dubs 'the display problem'.

2   Structural-connectives & star scheme

Display logic replaces Gentzen's polyvalent comma with a bivalent circle, $X \circ Y$. Like the comma, the $\circ$ means "something like" conjunction on the left and "something like" disjunction on the right of the turnstile. Display logic also has a single place star connective, $*X$, that allows one to flip structures from one side of the turnstile to the other and back again. Its meaning is often thought of as 'something like' negation. Although display logic has other structural connectives, this essay will focus on these two.

Strictly speaking, the generic un-indexed structural-connectives (like the un-indexed formula-connectives), are functions that map each family index (S4, r, e, h, b, etc.) into a specific structural connective of the family associated with that index. For the most part, however, I will suppress these indices and give generic un-indexed formulations of schemes of display equivalences. I hope the reader will not lose sight of them entirely, however, because much of the motivation for canvassing the possible schemes and their properties lies in the greater control this will afford logicians working in display logic over the basic properties they build into the connectives of their languages.

Every scheme I will deal with treats negative structuring, $*$, when it appears alone, in the same way. Star in each scheme has full contraposition and double star elimination. Consecutions on the same

---

[3]A more precise formulation of this property is that "each antecedent part X of a consecution S can be displayed as the antecedent (itself) of a display equivalent consecution $X \vdash W$; and the consequent W is determined only by the position of X in S, not by what X looks like. Similarly for consequent parts of S." (Anderson et al., 1992, p. 301)

line below are display equivalent (i.e. interderivable):

$$
\begin{array}{llll}
X \vdash Y & *Y \vdash *X & X \vdash **Y & **X \vdash Y \\
X \vdash *Y & Y \vdash *X & & \\
*X \vdash Y & *Y \vdash X & & \\
\ldots X \ldots & \ldots **X \ldots & &
\end{array}
$$

The last line is intended to convey general intersubstitutability, which follows from the first line only, in the presence of the other schemes granting the display property. We can think of the scheme as a set of display equivalence classes. Since schemes are sets of classes, in order to make things easier on the eye I will use square brackets to mark equivalence classes and reserve curly brackets for other sets. For example, leaving out the last line, the scheme in the above table is:

$$
\{ [X \vdash Y, \ *Y \vdash *X, \ X \vdash **Y, \ **X \vdash Y],
$$
$$
[X \vdash *Y, \ Y \vdash *X], [*X \vdash Y, \ *Y \vdash X] \}^4
$$

Consecutions of the form of one member of a class are interderivable with corresponding consecutions of the form of the other members of the class.

Although there is nothing essential about this treatment of negative structuring, having a connective that allows structure to be moved from one side of the turn-style to the other is of obvious use in securing the display property. In part this is because in every scheme (as well as every structural rule) antecedent [consequent] structures remain antecedent [consequent] parts, which guarantees that the same structure cannot be displayed both on the left and on the right.

3 The display problem

Because structural variables are schematic, restricting the components of the display equivalence classes only to antecedent parts does not result in a loss of generality, although it does prevent redundantly

---

[4]In what follows I will present the equivalence classes in set notation, but in order to make this notation more legible I will adopt Belnap's graphic method of presentation, in which consecutions on the same line are members of the same equivalence class. Although at the moment this may sound awkward, when it comes up below I think it will feel natural.

treating the same scheme in two different forms. To see this consider a consecution, $X \circ Y \vdash * Z$, composed entirely of antecedent parts and another version of the same consecution with consequent parts, e.g. $* X \circ Y \vdash * Z$. It is obvious that because each variable ranges over both X and $* X$, treating both consecutions in our schemes would lead to redundancy. As such, for simplicity and readability in the rest of the essay I will use only antecedent parts.[5]

What will differ between the schemes is how the star interacts with the binary circle. Since the largest arity connective in the formulation of display logic we are dealing with is binary, the schemes will at most contain three structural variables. Now given that I have proposed to deal only with star and circle and that the schemes for star alone, which involve only two variables, are presented above, what is left are the schemes involving three variables.

Formulating the consecutions only with antecedent parts, there are twelve that will be involved in the rest of our schemas. These are grouped by which variable is displayed:[6]

$Z(XY)$ group:	$X(YZ)$ group:	$Y(ZX)$ group:
(1) $Z \vdash * X \circ * Y$	(5) $X \vdash * Y \circ * Z$	(9) $Y \vdash * Z \circ * X$
(2) $Z \vdash * Y \circ * X$	(6) $X \vdash * Z \circ * Y$	(10) $Y \vdash * X \circ * Z$
(3) $Z \vdash *(X \circ Y)$	(7) $X \vdash *(Y \circ Z)$	(11) $Y \vdash *(Z \circ X)$
(4) $Z \vdash *(Y \circ X)$	(8) $X \vdash *(Z \circ Y)$	(12) $Y \vdash *(X \circ Z)$

We can think of these groups as sets.

With the exhaustion of the relevant consecutions in this list, the display problem becomes specified: each of the display equivalence classes in a scheme must have at least one member of each of the three groups.

An important related corollary of this condition that Belnap doesn't explicitly mention is that every complete scheme of display equivalences must include all twelve consecutions, although portions of the schemes can be largely independent of one another. This follows

---

[5]Belnap's reasons for treating only antecedent parts are different. (Belnap, 1996, p. 85)

[6]I have altered Belnap's numbering slightly because a more systematic numbering makes the relations between the schemes easier to notice. Belnap dubs these groups the $Z(XY)$, $X(YZ)$, $Y(ZX)$ 'families' (not to be confused with language families). To avoid the ambiguity, I'll instead use 'group'.

from the fact that if one of the twelve did not belong to an equivalence class, the display property would fail.

A re-lettering of a set of consecutions is obtained by exchanging every instance of a variable or variables with the same variable. Re-lettering will be an oft used strategy for figuring out how the schemes work and why the semi-lattice of schemes I am about to present is complete within the bounds set, so having a precise idea of it is important. Let $\alpha, \beta \in \{X, Y, Z\}$ and $\alpha \neq \beta$. Then an α/β re-lettering of a set of consecutions will involve exchanging all of the instances of α with instances of β and all of the instances of β with instances of α.

For example, the $Z(XY)$ group is an X/Z re-lettering of the $X(YZ)$ group. We can see this because exchanging X with Z in (1) yields (6), in (2) yields (5), in (3) yields (8), and in (4) yields (7). Similarly the $Z(XY)$ group is an Y/Z re-lettering of the $Y(ZX)$ group and the $X(YZ)$ group is an X/Y re-lettering of the $Y(ZX)$ group. When letters are not specified, any of the possible re-letterings will do.

Not including one of the consecutions in any display equivalence class would mean not including any of its re-letterings in an equivalence class. I explain this in detail in §6.

4   The schemes

In this section I will present all of the possible schemes. In the next I will discuss their properties. In §6 I will argue that this list is complete and that it forms a join semi-lattice. As above, all consecutions on the same line are in the same equivalence class.

There are three basic types of schemes. The first (the **GA, A, A', B, C_a** and **C_b** schemes) have three or six consecutions in each equivalence class. The second (the **P, P', Q** and **Q'** schemes) have four consecutions in each class. And the third type (which includes only the **easy** scheme) has all twelve consecutions in the same class.

All of the schemes of the first type are built out of the equivalence classes of the **GA** Scheme = $\{ GA_{1a}, GA_{1b}, GA_{2a}, GA_{2b} \}$

$GA_{1a} = [(3) \; Z \vdash *(X \circ Y), \quad (7) \; X \vdash *(Y \circ Z), \quad (11) \; Y \vdash *(Z \circ X)]$
$GA_{1b} = [(2) \; Z \vdash *Y \circ *X, \quad (6) \; X \vdash *Z \circ *Y, \quad (10) \; Y \vdash *X \circ *Z]$
$GA_{2a} = [(4) \; Z \vdash *(Y \circ X), \quad (8) \; X \vdash *(Z \circ Y), \quad (12) \; Y \vdash *(X \circ Z)]$
$GA_{2b} = [(1) \; Z \vdash *X \circ *Y, \quad (5) \; X \vdash *Y \circ *Z, \quad (9) \; Y \vdash *Z \circ *X]$

The equivalence classes of the **GA** scheme are the bases for three six-six schemes:

$$\mathbf{A}\text{ scheme} = \{\,A_1, A_2\,\} \qquad \mathbf{A'}\text{ scheme} = \{\,A_1', A_2'\,\}$$
$$A_1 = [GA_{1a} \cup GA_{1b}] \qquad A_1' = [GA_{1a} \cup GA_{2b}]$$
$$A_2 = [GA_{2a} \cup GA_{2b}] \qquad A_2' = [GA_{1b} \cup GA_{2a}]$$

$$\mathbf{B}\text{ scheme} = \{\,B_a, B_b\,\}$$
$$B_a = [GA_{1a} \cup GA_{2a}]$$
$$B_b = [GA_{1b} \cup GA_{2b}]$$

Two six-three-three schemes based on the equivalence classes of the **B** and **GA** schemes are also possible:

$$\mathbf{C_a}\text{ scheme} = \{\,B_a, GA_{1b}, GA_{2b}\,\} \quad \mathbf{C_b}\text{ scheme} = \{\,B_b, GA_{1a}, GA_{2a}\,\}$$

There are only four schemes of the second type:

$$\mathbf{P}\text{ scheme} = \{\,P_1, P_2, P_3\,\} \qquad \mathbf{P'}\text{ scheme} = \{\,P_1', P_2', P_3'\,\}$$
$$P_1 = [(3), (5), (6), (12)] \qquad P_1' = [(4), (5), (6), (11)]$$
$$P_2 = [(4), (7), (9), (10)] \qquad P_2' = [(3), (8), (9), (10)]$$
$$P_3 = [(1), (2), (8), (11)] \qquad P_3' = [(1), (2), (7), (12)]$$

$$\mathbf{Q}\text{ scheme} = \{\,Q_1, Q_2, Q_3\,\} \qquad \mathbf{Q'}\text{ scheme} = \{\,Q_1', Q_2', Q_3'\,\}$$
$$Q_1 = [(2), (7), \ (8), \ (9)] \qquad Q_1' = [(1), (7), \ (8), (10)]$$
$$Q_2 = [(1), (6), (11), (12)] \qquad Q_2' = [(2), (5), (11), (12)]$$
$$Q_3 = [(3), (4), \ (5), (10)] \qquad Q_3' = [(3), (4), \ (6), \ (9)]$$

The final scheme is just the **easy** scheme. It has one equivalence class that contains (1)–(12) so:

easy scheme $= \{[(1), (2), (3), (4), (5), (6), (7), (8), (9), (10), (11), (12)]\}$

5   Properties of the schemes

While I doubt all these schemes will be equally useful, I will prescind from passing judgment here on whether a scheme will likely find a use or not. Rather, my aim will be to point out what seem to be their interesting properties, which will help in my argument that the list of schemes is complete.

A basic feature that shapes the properties of the schemes of the twelve component consecutions is that they divide into those with an antecedent circle, and those with a consequent circle. If the circle occurs within an even [odd] number of stared parentheses on the left, then it is an antecedent [consequent] circle because when it is not contained within a star it will be on the left [right] side of the turn-style. As Belnap points out, the circle in the antecedent and circle in the consequent are independently specifiable connectives.[7] With the technique of re-lettering, we can explain this by noting that when we re-letter consecutions, an antecedent circle never becomes a consequent circle and vice versa.

We can cut back the number of schemes which deserve attention by pointing out that all of the prime schemes are only notational variants on the schemes of which they are the primes. We can easily define a circle governed by a primed scheme with a circle governed by the corresponding unprimed scheme, or vice versa. As Belnap points out in relation to the **A** and **A′** scheme, "one could obtain the **A′** scheme by starting out with the **A** scheme and defining a new operation $\circ'$ in consequent position by $X \circ' Y =_{\mathrm{df}} Y \circ X$".[8] The **Q′** scheme is derivable from the **Q** scheme in exactly the same way. And the **P′** scheme is derivable from the **P** scheme by defining a new operation $\circ'$ in antecedent position by $X \circ' Y = Y \circ X$.

This means that the basic distinction between the four-four-four schemes is whether the circle commutes in the antecedent (**Q** schemes) or consequent (**P** schemes).[9] These are also two important properties for the **GA** based schemes and seem to be two of the most potentially interesting properties the schemes can have. Neither of the **A** schemes have either property, nor does the **GA** scheme. But the $\mathbf{C}_a$ scheme commutes in the antecedent, while the $\mathbf{C}_b$ scheme commutes in the consequent, and the **B** scheme is the only scheme besides the **easy**

---

[7](Belnap, 1996, p. 89)

[8](Belnap, 1996, p. 90)

[9]The further distinction between the scheme and its prime describes how the $\circ$ interacts with negation. The **Q** scheme (**P** scheme) is distinguished from its prime by whether a structure being moved from the left to the right (right to left) side of the turn-style goes on the inside or outside of the previously displayed structure.

scheme that commutes in both.

Although the **A** schemes lack commutativity, if we wanted to press the truth-functional analogy, the equivalence classes of the **A** scheme allow something like a structural De Morgan's rule. That is, they postulate that a negated circle in the consequent disjoining two structures is equivalent to a circle in the antecedent conjoining those same structures negated.

The **GA** scheme has the fewest properties built into it: it neither commutes in the antecedent nor consequent, nor does it have the De Morgan-like property. As such, it allows the most control over the properties of the languages built using it. Nonetheless, the **GA** scheme and all of the schemes built out of its classes, share a common form or technique by which they can be constructed, that the **P** and **Q** schemes lack. Accordingly, there may be languages that can be built out of the latter schemes that cannot be built from the **GA** scheme.

The form or technique I have in mind is that each of the three members of any of the GA classes can be arrived at by taking one of the members, replacing X with Y, Y with Z, and Z with X twice, in order to get each of the other two members. (You can of course also go the other direction: replacing Y with X, Z with Y, and X with Z.) It is noteworthy that this technique of generating classes keeps the classes defining circle in the antecedent distinct from those defining circle in the consequent.

Similarly, all of the classes of **P** and **Q** schemes also share an underlying form or technique for construction. First, take any consecution and α/β re-letter it, for some α, β. This will be the second member of the class. For the third member pick a consecution displaying a variable not displayed by either of the other two, which involves the kind of circle that the other two do not (since all of these schemes inter-define the antecedent and consequent circle). To get the fourth and final member of the class, for the same α and β, α/β re-letter it. Once you have one class, the way to arrive at the other three classes in the scheme is by re-lettering this class twice (this will be explained further below).

The **easy** scheme has commutativity in the antecedent and consequent, and the De Morgan like property. Because of this, it is the scheme in which circle most closely approximates disjunction in the antecedent and conjunction in the consequent.

6 The completeness of the list

Although this is easy to verify through brute combinatorics, given that we only have twelve consecutions, in this section I will explain why this list exhausts the possible schemes using circle and star, while keeping the star scheme fixed. Before plunging in, it will help to note that because the letters are schematic we can actually express each of the schemes more simply because many of their equivalence classes are actually identical. We can see, for example, that the three equivalence classes of the **P** scheme are identical through re-lettering. An X/Y re-lettering of P_1 yields P_2, and vice versa. An X/Z re-lettering of P_1 yields P_3, and vice versa. A Y/Z re-lettering of P_2 yields P_3, and vice versa. A similar procedure can be followed with the rest of the **P** and **Q** schemes to show their classes are identical, and with the **GA** scheme to show GA_{1a} is identical with GA_{2a} and GA_{1b} is identical with GA_{2b}. As a result many of the schemes can be expressed more simply, if not more perspicuously as follows:

$$\textbf{GA} \text{ scheme} = \{\, GA_{1a}, GA_{1b}\,\} \qquad \textbf{A} \text{ scheme} = \{\, A_1\,\}$$
$$\textbf{P} \text{ scheme} = \{\, P_1\,\} \qquad \textbf{Q} \text{ scheme} = \{\, Q_1\,\}$$

(And similarly for the prime schemes.)

This fact is relevant here because it points to how inflexible the schemes are. If we switched some of the consecutions in one equivalence class with those of another in their scheme, then we would have to make corresponding switches in the other equivalence classes. Otherwise, by re-lettering the altered equivalence classes we would get an equivalence class that is for the most part identical with the corresponding unaltered one, except for the consecution that is the re-lettered newly introduced one. This consecution will belong to the other unaltered equivalence class, and so any consecution in either of the two unaltered equivalence classes will be inter-derivable from one another via the re-lettered altered equivalence class.

For example, suppose we switched (3) and (6) in the equivalence classes of the **GA** scheme to get the equivalence classes $GA_{1a}^+ =_{\mathrm{df}}$ $[(6), (7), (11)]$ and $GA_{1b}^+ =_{\mathrm{df}} [(2), (3), (10)]$. Now the X/Z re-lettering of GA_{1a}^+ has members of both GA_{2a} and GA_{2b}. Accordingly, using this re-lettering and these two classes, we can derive any member of GA_{2a} from a member of GA_{2b} and vice-versa. Thus, on the hypothesis

GA_{2a} and GA_{2b} collapse into A_2. Then, using a re-lettering of A_2, we can derive any member of GA_{1a}^+ from GA_{1b}^+ and vice-versa, so these collapse together into A_1 and we have the **A** scheme.

In general, the principle that this gives us is that the re-letterings of an equivalence class must be identical to another class (maybe themselves) already in the scheme. Otherwise, the scheme is unstable and we can use the re-lettered equivalence classes to collapse other equivalence classes together. With this principle what remains to be shown, in order to show the completeness of the list, is that these schemes are the only stable ones.

For this, it will help to have another principle governing the classes: if two members of a class are some kind of α/β re-lettering of one another, then there must be at least two other members of the class that are that same kind of re-lettering of one another. The schemes of the second type (**P**, **Q**, etc.) exhibit this rule, but now it can be shown as a corollary of the above principle. Take two consecutions that are some α/β re-lettering of one another. To these, at least one consecution displaying a not yet displayed variable must be added for the class to secure the display property. Now if we α/β re-letter all three members, the re-lettered first will be identical to the non-re-lettered second, and vice-versa, but the re-lettered third will be a new consecution that is an α/β re-lettering of the original third. But since the α/β re-lettering and the original proposed class share members, they collapse together by the above, and the principle is shown. For example, since (4) is a Y/Z re-lettering of (11) let $GA_{2a}^+ =_{\mathrm{df}} [(4),(8),(11)]$. Its Y/Z re-lettering will be: (4), (7), (11). Since (4) is shared, the two classes collapse together, and the original class must include (7) as well.

To see that the schemes canvassed are the only stable ones first note that no other schemes can be built from the classes of the **GA**, **P**, and **Q** schemes. Since there is only really one equivalence class in the **A**, **P**, and **Q** schemes adding an additional class to any of these would cause them to collapse into the easy scheme. Otherwise we have built all of the schemes it is possible to build from the **GA** scheme alone. This is because there are only really two distinct classes in that scheme, and the only scheme with more than one class that results from taking their unions is the **B** scheme, the classes of which are then used to build the **C** schemes.

Now the **GA** scheme and the schemes of the second type (**P**, **Q**, etc.) are minimal solutions[10] to the display problem in that there is no further refinements of either of them that also solve it. If there were other stable schemes then either they would have to be minimal solutions to the display problem, or be built by joining the classes of some other minimal solution to the display problem. So in order to show that the list of schemes presented is complete, it suffices to show there are no more minimal solutions to the display problem.

Since the circle in the antecedent or consequent can be defined independently, there will be those minimal schemes that define them separately (the **GA** scheme) and those that don't (the second type schemes). If a proposed new minimal solution defined them separately, then it would have to have at least two non-identical classes, dealing with the six consecutions governing each circle separately. These classes could either be composed of six or three members, because if they were composed of four or five they would be obviously unstable. If they were composed of all six consecutions, then they would not be minimal because they could be divided into the classes of the **GA** scheme. If they were composed of three member classes, however, then they must be those of the **GA** scheme because otherwise two of the members would be α/β re-letterings of one another, for some α and some β, which would entail that a fourth member belonged to the class by the above corollary, but this would cause collapse.

If the proposed minimal solution defined the two circles together, then it would have to include consecutions governing each circle in at least one of its classes. This class, like any, must solve the display problem, and so must have at least one member displaying each of the three variables. Either the third member of the class will be an α/β re-lettering of one of the other members or it will not be. If it is, then by the above corollary there will be a fourth member of the class which is an α/β re-lettering of the third, for the same α and β. Now since the class defines the antecedent and consequent circles simultaneously, it is obvious from this and the technique I gave for generating the schemes of the second type, that the possible classes which could ground those schemes are the same as the ones that satisfy this description, and all of the schemes that could be generated out of these have already been discussed.

If, however, the third member of the class is not an α/β re-lettering

---

[10]Borrowing the term from (Belnap, 1996, pp. 89–90).

of one of the other members, it at least must be the same kind of circle
(antecedent or consequent) as one of the other members. Since it is
not an α/β re-lettering of this consecution, however, of the other five
consecutions with this same kind of circle there are only two consecutions it could be, and it will be in the same class of the **GA** scheme
as this consecution. If we now re-letter this class in all three ways,
in each re-lettering we will have two out of the three consecutions in
the other GA class governing the type of circle that two of the three
members of the original class share, but which they are not themselves
in. That this would be so should be clear from the fact that no matter how one re-letters a member of the GA classes, one always gets a
member of the other GA class governing the same circle. From this
fact it also follows that the three re-letterings of the third member
of the class will all be different members of the same GA class. Now
since these three classes all share members, they will collapse together
and the resulting class will be the union of the two classes of the **GA**
scheme that the first three consecutions do not belong to (which will
be one of the **A** or **A'** scheme classes).

An example will help. Suppose the class we start from is:

$$GA_{2b}^* =_{df} [(1)\ Z \vdash * X \circ * Y,\ (5)\ X \vdash * Y \circ * Z,\ (11)\ Y \vdash *(X \circ Z)]$$

$X/Y, Y/Z$, and Z/X Re-lettering this we get:

$$GA_{1b}^{*XY} =_{df} [(2)\ Z \vdash * Y \circ * X,\ (7)\ X \vdash *(Y \circ Z),\ (10)\ Y \vdash * X \circ * Z]$$
$$GA_{1b}^{*YZ} =_{df} [(4)\ Z \vdash *(Y \circ X),\ (6)\ X \vdash * Z \circ * Y,\ (10)\ Y \vdash * X \circ * Z]$$
$$GA_{1b}^{*ZX} =_{df} [(2)\ Z \vdash * Y \circ * X,\ (6)\ X \vdash * Z \circ * Y,\ (12)\ Y \vdash *(X \circ Z)]$$

Now since GA_{1b}^{*XY} & GA_{1b}^{*YZ} share (10), and GA_{1b}^{*YZ} & GA_{1b}^{*ZX} share
(6) the three classes will collapse together, yielding A_2'.

Since the two classes of the **A'** (or **A**) scheme are identical by re-
lettering, this, or the **A** scheme, is the scheme we are committed to,
given that in our original class both circles are present and the third
member was not an α/β re-lettering of one of the other members. But
the **A** and **A'** schemes are not minimal solutions to the display problem, so there is no other minimal solution besides the ones canvassed.
Thus, the list of schemes given is complete.

Accordingly, this list forms a join semi-lattice on the twelve consecutions:

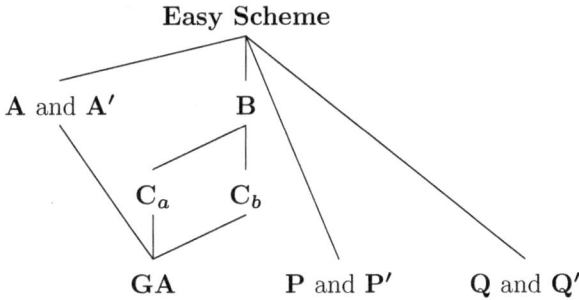

Easy Scheme

7   Conclusion

These properties of the behavior of the structural-connectives given the different schemes are, however, rather superficial. Since the languages formulable in display logic are individuated by the differences between the structural rules and display equivalences governing them, the deeper properties of the schemes (as well as the structural rules) are the ones they give the languages built using them. But so far I don't think anyone has a clear grasp on how each of the individual structural rules or schemes of equivalences effects the language families they are a part of or the extent to which the specific properties of the families can be traced back to individual rules. Although this paper has not tackled this difficult question, I hope that by providing a complete semi-lattice of display equivalence schemes for the standard connectives our understanding of the variety of structural resources within display logic for formulating interesting languages has been slightly improved.

References

Anderson, A., Jr, N. B., & Dunn, J. (1992). *Entailment. the logic of relevance and necessity* (Vol. 2). Princeton: Princeton University Press.

Belnap, N. (1982). Display logic. *Journal of Philosophical Logic,* *11*(4), 375–417.

Belnap, N. (1996). The display problem. In H. Wansing (Ed.), *Proof theory of modal logic* (pp. 79–92). Dordrecht, Boston, and London: Kluwer Academic Publishers.

Gentzen, G. (1969). Investigations into logical deduction. In M. E. Szabo (Ed.), *The Collected Works of Gerhard Gentzen* (pp. 68–131). Amsterdam: North-Holland Publishing Company.

Tyke Nunez
Department of Philosophy
University of Pittsburgh
1001 Cathedral of Learning
Pittsburgh, PA 15260
e-mail: asn13@pitt.edu

Logic as Based on Incompatibility

Jaroslav Peregrin[*]

Abstract

The aim of the paper is to tackle two related questions: *Is it possible to reduce the foundations of logic to the mere concept of incompatibility?* and *Does this reduction lead us to a specific logical system?* We conclude that the answers, respectively, are YES and a qualified NO (qualified in the sense that basing semantics on incompatibility does make some logical systems more natural than others, but without ruling out the alternatives.)

1 Can inference serve as a foundation of logic?

Can we base the whole of logic solely on the concept of incompatibility? My motivation for asking this is two-fold: firstly, a technical interest in what a minimal foundations of logic might be; and secondly, the existence of philosophers who have taken incompatibility as the ultimate key to human reason (*viz.*, e.g., Hegel's concept of *determinate negation*). The main aim of this contribution is to tackle two related questions: *Is it possible to reduce the foundations of logic to the mere concept of incompatibility?* and *Does this reduction lead us to a specific logical system?* We conclude that the answers, respectively, are YES and a qualified NO (qualified in the sense that basing semantics on incompatibility does make some logical systems more natural than others, but without ruling out the alternatives.

A search for the bare bones of logic generally leads one to the relation of inference (or consequence). This way is explored meticulously by Koslow (1992). He defines an *implication structure* as, in effect,

---

[*]Work on this paper was supported by the research grant No. P401/10/1279 of the Czech Science Foundation.

an ordered pair $\langle S, \vdash \rangle$, where S is a set and $\vdash \subseteq \mathrm{Pow}(S) \times S$ fulfilling certain (relatively simple) restrictions. And obviously if we reduce incompatibility to inference, which is achievable by the well known *ex contradictione quodlibet* principle, we reach a logic based on incompatibility. The kind of logic flowing most straightforwardly from this setting is the intuitionist one.

However, there is also the approach taken by Brandom and Aker (2008), who have set up a logic based directly on incompatibility. They define an incompatibility structure as an ordered pair $\langle S, \perp \rangle$ such that S is a set and $\perp \subseteq \mathrm{Pow}(S)$ (again fulfilling certain restrictions). The authors introduce logical operators in such a way that they reach classical logic.

Does this mean that inference 'naturally' leads to intuitionist logic, whereas incompatibility leads to the classical one? Myself, I have argued that it is indeed intuitionist logic that is *the* logic of inference (see Peregrin, 2008). However, this should not be read as saying that choosing inference as the fundamental logical notion predetermines us to have intuitionist logic, and that choosing incompatibility as such a notion perhaps predetermines us to have the classical one. I will argue instead that we can use incompatibility to lay the foundation of almost any imaginable kind of logic.

In what follows I first analyze the relationship between the two above mentioned approaches to logic based upon incompatibility, and the source of the difference between the ensuing logics; with the result that the source of the difference is not the choice of the fundamental notion itself, but rather certain collateral desiderata that Koslow, but not Brandom and Aker, poses for his logical system. Then I indicate that by generalizing the approach of Brandom and Aker, in a variety of ways, we can also reach other kinds of non-classical logic.

2   The framework

A *generalized inferential structure* (*gis*) will be the ordered triple $\langle S, \perp, \vdash \rangle$, where S is a set, $\perp \subseteq \mathrm{Pow}(S)$ and $\vdash \subseteq \mathrm{Pow}(S) \times S$. Which constraints should be placed on the notions of incompatibility and inference on this maximally general level? (It is clear that not any kind of set of sets of sentences can be reasonably seen as instantiating incompatibility, and that not every relation between sets of sentences and sentences can reasonably be called a relation of inference.)

Before introducing the constraints, a word about notation. The variables X, Y, Z will range over subsets of S, whereas the variables A, B, C will range over elements of S. $\perp X$ will denote that $X \in \perp$. $X \vdash A$ will denote that $\langle X, A \rangle \in \vdash$. We will write X, Y after $\perp$ or before $\vdash$ as a shortcut for $X \cup Y$, and A as a shortcut for $\{A\}$. Hence, e.g. $\perp X, A$ expands to $X \cup \{A\} \in \perp$. Now we can list the constraints we consider basic:

($\perp$) for every X, Y: if $\perp X$ and $X \subseteq Y$, then $\perp Y$

($\vdash 1$) for every X, A: $X, A \vdash A$

($\vdash 2$) for every X, Y, A, B: if $X, A \vdash B$ and $Y \vdash A$, then $X, Y \vdash B$

Let us adopt a further notational convention. Symbols that appear as "free" in the conditions of the above kind will be understood as universally quantified. Given this convention, we can shorten the above conditions to:

($\perp$) if $\perp X$ and $X \subseteq Y$, then $\perp Y$

($\vdash 1$) $A, X \vdash A$

($\vdash 2$) if $X, A \vdash B$ and $Y \vdash A$, then $X, Y \vdash B$

($\perp$) states that an incompatible set of sentences cannot be turned into a compatible one by addition of further sentences. This is the single constraint stipulated by Brandom and Aker. ($\vdash 1$) states that if A belongs to X, then it is entailed by X. ($\vdash 2$) says that the relation of consequence is transitive (if X entails every element of a set that entails A, then X entails A). The constraints ($\vdash 1$) and ($\vdash 2$) are stipulated by Koslow; they are tantamount to the so-called Gentzenian structural rules.[1]

It is sometimes useful to replace ($\vdash 1$) by two other conditions:

---

[1]Gentzen (1934, 1935) introduced structural rules with which to characterize those relations of inference that he took to be 'standard'. In a slightly more contemporary manner, they can be summarized as follows:

$A \vdash A$	(*reflectivity*)
if $X, Y \vdash A$, then $X, B, Y \vdash A$	(*weakening* or *extension*)
if $X, A, A, Y \vdash B$, then $X, A, Y \vdash B$	(*contraction*)
if $X, A, B, Y \vdash C$, then $X, B, A, Y \vdash C$	(*permutation* or *exchange*)

Lemma 1 *Let $\langle S, \perp, \vdash \rangle$ be a gis for which $(\vdash 2)$ holds. Then $(\vdash 1)$ holds if and only if $(\vdash 1.1)$ and $(\vdash 1.2)$ hold:*

$(\vdash 1.1)$ $A \vdash A$;

$(\vdash 1.2)$ *if $X \vdash A$, then $X, B \vdash A$.*

Proof. That $(\vdash 1)$ follows from $(\vdash 1.1)$ and $(\vdash 1.2)$ is obvious; conversely it is obvious that $(\vdash 1.1)$ follows from $(\vdash 1)$; so the only thing to show is that $(\vdash 1.2)$ follows from $(\vdash 1)$. Assume $X \vdash A$. As X, B entails all elements of X, $X, B \vdash A$ is yielded by the repeated application of $(\vdash 2)$. $\square$

We may further consider constraints on the interplay of $\vdash$ and $\perp$. The most natural ones seem to be the following two:

$(\perp\vdash 1)$ if $\perp X$, then $X \vdash A$

$(\vdash\perp 1)$ if $X \vdash A$ and $\perp Y, A$, then $\perp Y, X$

The first says that an incompatible set entails everything (a version of the famous *ex falso quodlibet*, or, perhaps better, *ex contradictione quodlibet*); the second says that whatever is compatible with the antecedent of a consequence, cannot be incompatible with its consequent (Brandom and Aker call this condition—more precisely an equivalent one—*defeasibility*).

Adopting either of these conditions together with its converse allows us to reduce one of the two basic concepts to the other:

$(\perp\vdash 2)$ if $X \vdash A$ for every A, then $\perp X$

$(\vdash\perp 2)$ if $\perp Y, A$ implies $\perp Y, X$ for every Y, then $X \vdash A$

Thus, adopting $(\perp\vdash 1)$ plus $(\perp\vdash 2)$ is tantamount to reducing $\perp$ to $\vdash$, as treating a set of sentences as incompatible just in the case that it entails everything. From the other side, adopting $(\vdash\perp 1)$ plus $(\vdash\perp 2)$ is tantamount to reducing $\vdash$ to $\perp$, as treating a sentence as inferable

---

if $X, A, Y \vdash B$ and $Z \vdash A$, then $X, Z, Y \vdash B$ (*cut*)

Within our framework, two of the conditions, namely *contraction* and *permutation*, are implicit to the assumption that inference is a relation between *sets* (rather than sequences) of sentences and sentences.

from a set of sentences just in the case that whatever is incompatible with the consequent is incompatible with its antecedent.

Let us call a gis *standard* iff it complies with ($\vdash$1), ($\vdash$2), ($\bot$), ($\vdash\bot$1), ($\vdash\bot$2), ($\bot\vdash$1), and ($\bot\vdash$2). Some of these constraints turn out to be superfluous.

Lemma 2 *Let* $\langle S, \bot, \vdash \rangle$ *be a gis for which* ($\bot\vdash$1), ($\bot\vdash$2), ($\vdash$2), *and* ($\vdash\bot$2) *hold. Then* ($\vdash\bot$1) *holds.*

Proof. Assume that $X \vdash A$ and $\bot\, Y, A$. Given ($\bot\vdash$1), it follows $Y, A \vdash B$ for every B. Given ($\vdash$2) we get $Y, X \vdash B$ for every B, and finally using ($\bot\vdash$2), we reach $\bot\, Y, X$. □

The situation is similar with ($\bot\vdash$1):

Lemma 3 *Let* $\langle S, \bot, \vdash \rangle$ *be a gis for which* ($\bot$) *and* ($\vdash\bot$2) *hold. Then* ($\bot\vdash$1) *holds.*

Proof. Assume that $\bot\, X$. It follows from ($\bot$) that $\bot\, X, Y$ for every Y, hence it holds trivially that for every Y, A, if $\bot\, Y, A$, then $\bot\, Y, X$. Given ($\vdash\bot$2), $X \vdash A$. □

Corollary 4 *Let* $\langle S, \bot, \vdash \rangle$ *be a gis for which* ($\bot$), ($\bot\vdash$2), ($\vdash$2), *and* ($\vdash\bot$2) *hold. Then both* ($\vdash\bot$1) *and* ($\bot\vdash$1) *hold.*

Corollary 5 $\langle S, \bot, \vdash \rangle$ *is standard iff it complies with* ($\vdash$1), ($\vdash$2), ($\bot$), ($\vdash\bot$2), *and* ($\bot\vdash$2).

Lemma 6 *Let* $\langle S, \bot, \vdash \rangle$ *be a gis for which* ($\vdash\bot$1) *and* ($\vdash\bot$2) *hold. Then* ($\vdash$1.1) *and* ($\vdash$2) *hold; and* ($\vdash$1.2) *holds if* ($\bot$) *holds.*

Proof. As $\bot\, Y, A$ trivially implies $\bot\, Y, A$ for every Y and A, ($\vdash$1.1) holds for every A. To show that ($\vdash$2) holds, assume that $X, A \vdash B$ and $Y \vdash A$. In force of ($\vdash\bot$1), $\bot\, Z, B$ implies $\bot\, Z, X, A$ for every Z and $\bot\, Z, A$ implies $\bot\, Z, Y$ for every Z, hence especially $\bot\, Z, X, A$ implies $\bot\, Z, X, Y$ for every Z; and thus $\bot\, Z, B$ implies $\bot\, Z, X, Y$ for every Z. Hence, in force of ($\vdash\bot$2), $X, Y \vdash B$. Now assume that ($\bot$) holds and that $X \vdash A$. According to ($\vdash\bot$1), $\bot\, Z, A$ implies $\bot\, Z, X$ for every Z, and hence it also implies $\bot\, Z, X, B$ for every Z. Hence $X, B \vdash A$. □

Lemma 7 *Let* $\langle S, \bot, \vdash \rangle$ *be a gis for which* ($\bot\vdash$1) *and* ($\bot\vdash$2) *hold. Then* ($\bot$) *holds if* ($\vdash$1.2) *holds.*

Proof. Assume ($\vdash$1.2) and assume $\bot\, X$ and $X \subseteq Y$. Then, according to ($\bot\vdash$1), $X \vdash A$ for every A, and hence, according to ($\vdash$1.2), $Y \vdash A$ for every A. Hence according to ($\bot\vdash$2), $\bot\, Y$. $\qquad\qquad$ □

Corollary 8 *Let $\langle S, \bot, \vdash \rangle$ be a gis for which* ($\bot\vdash$1), ($\bot\vdash$2), ($\vdash\bot$1), ($\vdash\bot$2) *hold. Then* ($\bot$) *holds iff* ($\vdash$1.2) *holds.*

Given a gis, we can also define a *new* variant of incompatibility in terms of inference; and a new variant of inference in terms of incompatibility:

$$\triangle X =_{\text{Def.}} \text{ for every } A, X \vdash A$$

$$X \triangleright A =_{\text{Def.}} \text{ for every } Y, \text{ if } \bot\, Y, A \text{ then } \bot\, Y, X$$

For a general gis, there is, of course, no guarantee that these new versions will coincide (or even be similar to) the original ones. However, suppose that the $\vdash$ which serves as the basis for the definition of $\triangle$ is already reducible to the original $\bot$—hence suppose that ($\vdash\bot$1) and ($\vdash\bot$2) hold. Then the definition of $\triangle$ in terms of $\vdash$ as if 'undoes' the definition of $\vdash$; and the two pairs of operators coincide.

This means that given ($\vdash\bot$1) and ($\vdash\bot$2), $\triangle$ (trivially) coincides with $\bot$; and analogously for $\triangleright$ and $\vdash$. Combining this trivial result with the claim of Corollary 4, we get:

Theorem 9[2] *Let $\langle S, \bot, \vdash \rangle$ be a gis for which* ($\bot$), ($\bot\vdash$2), ($\vdash$2), *and* ($\vdash\bot$2) *hold. Let $\triangle$ and $\triangleright$ be defined as above. Then $\triangle$ coincides with $\bot$ and $\triangleright$ coincides with $\vdash$.*

In this case we can say that $\bot$ and $\vdash$ 'fit together' in the sense that one can be reconstructed from the other.

3   Incompatibility vs. inference

Let us now turn our attention to logical operators; we will restrict ourselves just to two of them, namely negation and conjunction.

It seems that the minimal requirements that must be put on negation are the following:

---

[2] Half of this theorem was proved by Brandom and Aker under the name of the *Representation theorem (of consequence relations by incompatibility relations)*. What they proved was that $\triangleright$ coincides with $\vdash$, given ($\vdash$2) and ($\vdash\bot$2), which they call, respectively, *general transitivity* and *defeasibility*.

($\neg K1$) $\perp A, \neg A$

($\neg K2$) if $\perp A, B$, then $B \vdash \neg A$

These constraints stipulate that negation of A is its *minimal incompatible*: ($\neg 1$) states that $\neg A$ is incompatible with A, whereas ($\neg 2$) states that any other sentence incompatible with A is inferable from B. If we reduce incompatibility to inference, i.e. accept ($\perp\vdash 1$) and ($\perp\vdash 2$), we get:

($\neg K1'$) $A, \neg A \vdash B$

($\neg K2'$) if $A, B \vdash C$ for every C, then $B \vdash \neg A$

This gives us a natural characterization of negation in terms of inference. The consequences of this stipulation were investigated by Koslow; it leads to the intuitionist kind of negation.

Let us now consider a slight generalization of Koslow's definition more suitable for our purposes (the first condition stays the same):

($\neg 1$) $\perp A, \neg A$

($\neg 2$) if $\perp A, X$, then $X \vdash \neg A$

In his exposition of the character of physical laws, Feynman (1985) indicates how physical laws can be brought to the common denominator of a *minimum principle*; and the Koslowian approach to logic can be seen as the way of reducing logical operators (and the laws governing them) to a similar common denominator: all of them mark minima and maxima of functions defined in terms of inference. (This is not Koslow's invention, it goes back to Hertz and Gentzen, but Koslow has treated it more systematically.) However, if we now abandon this program and simply seek for any reasonable assortment of rules constitutive of negation in terms of inference (plus possibly incompatibility), we can think about a constraint that is dual to ($\neg 2'$):

($\neg 3$) if $\perp \neg A, X$, then $X \vdash A$

This constraint stipulates that the negation of A is a sentence whose *minimal incompatible* is A. It is important that this stipulation entails the law of double negation, which distinguishes classical from intuitionist logic.

Lemma 10 *Let $\langle S, \perp, \vdash \rangle$ be a gis for which $(\neg 1)$ and $(\neg 3)$ hold. Then $\neg\neg A \vdash A$ for every A.*

Proof. According to $(\neg 3)$ it is the case that if $\neg A \perp \neg\neg A$, then $\neg\neg A \vdash A$; and $\neg A \perp \neg\neg A$ is an instance of $(\neg 1)$.   □

Brandom and Aker's definition of negation in terms of incompatibility is the following:

   $(\neg B1)$ if $\perp X, \neg A$, then $X \vdash A$;

   $(\neg B2)$ if $X \vdash A$, then $\perp X, \neg A$,

where $\vdash$ serves as a shortcut for "for every $Y \subseteq S$, if $\perp Y, A$, then $\perp Y, X$"; in other words we assume $(\vdash\perp 1)$ and $(\vdash\perp 2)$. What is the relationship between this definition and the above one? The answer is provided by the following two theorems:

Theorem 11 *Let $\langle S, \perp, \vdash \rangle$ be a gis for which $(\vdash\perp 1)$, $(\neg 1)$, and $(\neg 3)$ hold. Then $(\neg B1)$ and $(\neg B2)$ hold.*

Proof. As $(\neg B1)$ coincides with $(\neg 3)$, the only thing to prove is $(\neg B2)$. Hence assume $X \vdash A$. Then, according to $(\vdash\perp 1)$, it is the case that for every Y it holds that if $\perp Y, A$, then $\perp Y, X$. Hence especially if $\perp \neg A, A$, then $\perp \neg A, X$. Hence given $(\neg 1)$, we have $\perp \neg A, X$.   □

Theorem 12 *Let $\langle S, \perp, \vdash \rangle$ be a gis for which $(\vdash 1)$, $(\perp 1)$, $(\perp\vdash 1)$, $(\perp\vdash 2)$, $(\neg B1)$ and $(\neg B2)$ hold. Then also $(\neg 1)$, $(\neg 2)$, and $(\neg 3)$ hold.*

Proof. We have already noted that $(\neg 3)$ coincides with $(\neg B1)$; and as $(\vdash 1)$ implies $A \vdash A$, $(\neg B2)$ yields us $(\neg 1)$; hence the only thing to prove is $(\neg 2)$. To prove it, assume that $\perp A, B$ and further assume that $\perp Y, \neg A$. Then, in force of $(\perp 1)$, $\perp A, B, Y$, and, in force of $(\neg B1)$, $Y \vdash A$. Thus, in force of $(\vdash\perp 1)$, $\perp Y, B, Y$, i.e. $\perp Y, B$. Hence for any arbitrary Y, if $\perp Y, \neg A$, then $\perp Y, B$, which yields $X \vdash \neg A$ via $(\vdash\perp 1)$.   □

It follows that in a standard gis, $(\neg B1)$ and $(\neg B2)$ are equivalent to $(\neg 1)$, $(\neg 2)$ and $(\neg 3)$. Hence the fact that Koslow's approach leads to intuitionist negation, whereas Brandom's leads to the classical one, does not mirror any inherent difference between inference and incompatibility; the two approaches diverge because Koslow (like Gentzen) lays down specific constraints on how a feasible inferential definition of a logical constant ought to look, and that these constraints are fulfilled by (¬1) plus (¬2), not, however, by (¬1), (¬2) and (¬3).

Now let us turn our attention to conjunction. Koslow's way of introducing it is as the upper bound (or supremum, if you prefer):

($\wedge K1$) $A \wedge B \vdash A$

($\wedge K2$) $A \wedge B \vdash B$

($\wedge K3$) if $X \vdash A$ and $X \vdash B$, then $X \vdash A \wedge B$

Lemma 13 *Let $\langle S, \perp, \vdash \rangle$ be a gis for which $(\wedge K1)$, $(\wedge K2)$, $(\vdash 1)$, $(\vdash 2)$ hold. Then $(\wedge K3)$ is equivalent to*

($\wedge K3'$) $A, B \vdash A \wedge B$.

Proof. Assume ($\wedge K3$). As ($\vdash 1$) yields us $A, B \vdash A$ and $A, B \vdash B$, we get $A, B \vdash A \wedge B$ by means of ($\vdash 2$). Now assume ($\wedge K3'$) and assume $X \vdash A$ and $X \vdash B$. We get ($\wedge K3$) by means of ($\vdash 2$). $\square$

Brandom and Aker's definition of conjunction is the following:

($\wedge B1$) if $\perp X, A \wedge B$, then $\perp X, A, B$

($\wedge B2$) if $\perp X, A, B$, then $\perp X, A \wedge B$

Theorem 14 *Let $\langle S, \perp, \vdash \rangle$ be a gis for which $(\wedge K1)$, $(\wedge K2)$, $(\wedge K3)$, and $(\vdash \perp 1)$ hold. Then $(\wedge B1)$ and $(\wedge B2)$ hold.*

Proof. To prove ($\wedge B1$), assume $X \perp A \wedge B$. We obtain $\perp X, A, B$ by means of ($\vdash \perp 1$) and ($\wedge K3$). To prove ($\wedge B2$), assume $\perp X, A, B$. Using ($\vdash \perp 1$) and ($\wedge K1$) we obtain $\perp X, B, A \wedge B$; and using ($\vdash \perp 1$) and ($\wedge K2$) we further obtain $\perp X, A \wedge B, A \wedge B$; hence $\perp X, A \wedge B$. $\square$

Theorem 15 *Let $\langle S, \perp, \vdash \rangle$ be a gis for which $(\wedge B1)$, $(\wedge B2)$, $(\perp 1)$, $(\vdash \perp 2)$ hold. Then $(\wedge K1)$, $(\wedge K2)$, and $(\wedge K3)$ hold.*

Proof. To prove ($\wedge K1$), assume $\perp X, A$. Using ($\perp 1$) we get $\perp X, A, B$. ($\wedge B2$) then yields us $\perp X, A \wedge B$. Hence if $\perp X, A$, then $\perp X, A \wedge B$; and using ($\vdash \perp 2$) we get ($\wedge K1$). The proof of ($\wedge K2$) is straightforwardly analogous. To prove ($\wedge K3$), assume $\perp X, A \wedge B$. With the help of ($\wedge B1$) we obtain $\perp X, A, B$. Now we get $A, B \vdash A \wedge B$ using ($\vdash \perp 2$). $\square$

This ultimately clarifies the relationship between Koslow's inferential logic and Brandom and Aker's incompatibility logic, and explains why the former leads to the intuitionist, while the latter to classical logic. The distinctness of the outcomes do not stem from any inherent differences between inference and incompatibility (indeed the two concepts are two sides of the same coin); rather, it stems from the discipline that Koslow, in contrast to Brandom and Aker, adds to his inferential foundations.

4   Beyond classical and intuitionist logic

We have seen that basing logic on incompatibility and/or inference naturally leads us to either classical, or intuitionist logic. Is it possible to conceive of basing other kinds of logic, such as relevant logic or modal logic, on incompatibility and/or inference? The case of relevant logic is straightforward: we know that to be able to introduce it, we need first to eliminate:

($\vdash$1.2) if $X \vdash A$, then $X, B \vdash A$,

which, as we saw in Corollary 8, is equivalent to ($\perp$). Hence all it takes to prepare the ground for the relevantist version of logic within the framework of incompatibility logic is to retract ($\perp$).

To accommodate other kinds of substructural logics, such as linear logic, we must interfere deeper with our foundations. It is well known that linear logic requires us to see inference as a relation not between subsets of S and elements of S, but rather between multisets of the elements of S and elements of S, which allows us to discard the structural rule of contraction. And we can go further and replace multisets with sequences, which gives us the opportunity to discard permutation, thus allowing for logics for which the order of premises—and, as the case may be, conclusions—is significant, e.g. some dynamic logics. The situation is much the same as for incompatibility logic: to make room for linear logic we must consider incompatibility not as a property of sets, but rather of multisets (so that, for example, the multiset $\{A, A, B\}$ may be inconsistent though $\{A, B\}$ is consistent; or *vice versa*) and we might further consider it as a property of sequences.

Of course, there is a conceptual question concerning the extent to which it makes sense to consider incompatibility as a property of sequences of sentences rather than its sets (just as there is the conceptual question of how far it makes sense to consider inference as a relation between sequences of sentences, rather than between sets of sentences and sentences). But at least some reasons appear to be available: we know, for example, that a collection of sentences presented in one order may make up a consistent story, whereas the same sentences in a different order may not.

How is it with modal logic? Brandom and Aker introduced a natural definition of the necessity operator based on incompatibility:

($\Box$) $\perp X, \Box A$ iff $\perp X$ or there is an Y such that not $\perp X, Y$ and not $Y \vdash A$.

They show that this definition of incompatibility leads to the simplest modal logic S5. Can we have different modal logics based on incompatibility? In principle, surely we can, provided that we add some surplus ingredient corresponding to the relation of equivalence on Kripkean models. Elsewhere (see Peregrin, 2010) I have shown how we may reach logic B in this way.

5 Conclusion

It is possible to base logic solely on the concept of incompatibility; and in fact it does not restrict us in any substantial way w.r.t. the kind of logic we want to have. Brandom and Aker's elegant way of establishing it leads us to classical logic and further possibly to S5. (In contrast to logic based on inference that appears to naturally yield intuitionist logic.) In this sense, these logics may appear to be "natural" from the viewpoint of incompatibility logic; however, if naturalness is not what we are after, nothing prevents us from erecting almost any kind of logic on incompatibility foundations.

References

Brandom, R., & Aker, A. (2008). Between Saying and Doing: Towards Analytical Pragmatism. In R. Brandom (Ed.), (chap. Appendix to Chapter V). New York: Oxford University Press.

Feynman, R. (1985). *The Character of Physical Laws*. Cambridge (Mass.): MIT Press.

Gentzen, G. (1934). Untersuchungen über das logische Schliessen I. *Mathematische Zeitschrift*, *39*, 176–210.

Gentzen, G. (1935). Untersuchungen über das logische Schliessen II. *Mathematische Zeitschrift*, *39*, 405–431.

Koslow, A. (1992). *A Structuralist Theory of Logic*. Cambridge: Cambridge University Press.

Peregrin, J. (2008). What is *the* logic of inference? *Studia Logica*, *88*, 263–294.

Peregrin, J. (2010). Brandom's Incompatibility Semantics. *Philosophical Topics*, *36*, 99–122.

Jaroslav Peregrin
Department of Logic
Institute of Philosophy
Academy of Scicences of the Czech Republic
Jilská 1, 110 00 Prague 1
Czech Republic
e-mail: jarda@peregrin.cz
URL: http://jarda.peregrin.cz

The Logic of Falsification

Andreas Pietz[*]

Abstract

In this essay we will suggest a logic of falsification. This logic is supposed to capture certain ideas about semantics put forth by Michael Dummett. The basic idea is that assertions may be take to be correct as long as they cannot be falsified. The logic thus treats non-falsifiability as the property to be transferred from premises to conclusions. We will characterize the logic semantically and discuss a problem that arises with respect to incoherent statements.

1   Introduction

Michael Dummett has in several places claimed that falsification is the central concept to build a semantic theory around. Many will be surprised to see this claim associated with Dummett, who is after all known as an ardent defender of a *verificationistic* theory of meaning. However, at some prominent points he goes against verificationism and offers arguments like these:

> By making an assertion, a speaker rules out certain possibilities; (...) Thus, in the order of explanation, the notion of the incorrectness of an assertion is prior to that of its correctness. (...) These considerations prompt the construction of a different theory of meaning, one which agrees with the verificationist theory in making use only

---

[*]The research for this paper was partly funded by grant HUM2008-FFI04263 awarded by the Spanish Ministerio de Ciencia e Innovacin. I'd like to thank Graham Priest, Heinrich Wansing, Genoveva Marti, Sven Rosenkranz, Roberto Ciuni and the audiences at Logica in Hejnice and LLR10 in Ghent for helpful comments.

of effective rather than transcendental notions, but which
replaces verification by falsification as the central notion
of the theory: we know the meaning of a sentence when
we know how to recognize that it has been falsified.[1]

The idea is that an assertion is correct if and only if it is not incorrect, that is, if and only if it can not be falsified. Here, there is not
enough space to discuss this norm and its plausibility in Dummett's
programme much further. For a scenario where such an account will
not need much by way of motivation, you may want to think about a
criminal trial and especially the role of the defendant. As such trials
are conducted under the *presumption of innocence*, the assertions that
the defendant makes will indeed be taken to be correct as long as the
prosecution cannot prove them wrong.

2   Verification and falsification conditions

The next step is to define a notion of logical consequence and to spell
out the meaning of the logical constants. Given that logic will license
inferences from correctly assertable statements to other correctly assertable statements, we see that the property that should be transmitted from premises to conclusions is non-falsifiability. Dummett made
an ill-fated attempt at constructing such a logic,[2] but never returned
to the idea. The problem was that he thought that such a logic should
only make reference to falsifications and had no room for verifications,
and the resulting logic came out somewhat unbalanced. We will instead employ both verifications and falsifications to obtain what we
feel is a much more attractive logic.[3]

The model for our proposal will be the interpretation of intuitionistic logic. Whereas in the famous BHK-interpretation, the meanings
of the logical constants are given in terms of proof/verification conditions, we will give an account in terms of both verification and
falsification conditions. First, we should demand that no statement
should be both verified and falsified. Then, we should think about
verification and falsification conditions for compound statements.

---

[1](Dummett, 1993)

[2](Dummett, 1993, p. 83)

[3]And also a logic that fits much better to what Dummett says in the essay in
question, though again there will not be space for such exegetical arguments.

It is very plausible to say that a conjunction is falsified by a falsification of *either* conjunct and a disjunction by a falsification of the first *and* a falsification of the second disjunct.

If verifications and falsifications are around, the most natural role for negation to play is as a toggle between them: A negation is verified (falsified) iff the negated statement iff falsified (verified).

The conditional, as Dummett also notes in the essay, has a very clear falsification condition as well: The antecedent has to be verified, the succedent falsified. For the verification condition, the question is not all that clear, but the understanding of intuitionistic logic seemed to give a good approximation.

We can collect and express the preceding thoughts in a list of BHK-style clauses:

- c is a verification of $A \wedge B$ iff c is a pair (c_1, c_2) such that c_1 is a verification of A and c_2 is a verification of B

- c is a falsification of $A \wedge B$ iff c is a pair (i, c_1) such that $i = 0$ and c_1 is a falsification of A or $i = 1$ and c_1 is a falsification of B

- c is a verification of $A \vee B$ iff c is a pair (i, c_1) such that $i = 0$ and c_1 is a verification of A or $i = 1$ and c_1 is a verification of B

- c is a falsification of $A \vee B$ iff c is a pair (c_1, c_2) such that c_1 is a falsification of A and c_2 is a falsification of B

- c is a verification of $A \supset B$ iff c is a procedure that converts each verification d of A into a verification $c(d)$ of B

- c is a falsification of $A \supset B$ iff c is a pair (c_1, c_2) such that c_1 is a verification of A and c_2 is a falsification of B

- c is a verification of $-A$ iff c is a falsification of A

- c is a falsification of $-A$ iff c is a verification of A

3   Nelson logic

These clauses (together with the demand that nothing should be both verified and falsified) correspond to the Nelson logic N_3 (cf. López-Escobar, 1972 and Wansing, 1993). We will give a semantics for this logic that works similarly to the Kripke semantics for intuitionistic logic.

A model for N_3 is a structure $[W, \leq, v]$, W being a set of partially ordered ($\leq$) worlds or information states and v a valuation function from formulas to 1 and 0. Unlike in Kripke-models for intuitionistic logic, we allow v to be a *partial* function.[4] For $v_w(p) = 1$ I will write $w \Vdash_1 p$, and for $v_w(p) = 0$ I will write $w \Vdash_0 p$, leaving the relativization to v implicit. $w \Vdash_1 p$ means that p is verified at world w, $w \Vdash_0 p$ that p is falsified at that world.

We will have hereditary constraints for both 1 and 0:

For all p, and all worlds w and w',

$$\text{if } w \leq w' \text{ and } w \Vdash_1 p, \text{ then } w' \Vdash_1 p, \text{ and}$$

for all p, and all worlds w and w',

$$\text{if } w \leq w' \text{ and } w \Vdash_0 p, \text{ then } w' \Vdash_0 p.$$

We have to give separate clauses for $\Vdash_1$ and $\Vdash_0$ when defining the connectives:

$$w \Vdash_1 A \wedge B \text{ iff } w \Vdash_1 A \text{ and } w \Vdash_1 B$$
$$w \Vdash_0 A \wedge B \text{ iff } w \Vdash_0 A \text{ or } w \Vdash_0 B$$
$$w \Vdash_1 A \vee B \text{ iff } w \Vdash_1 A \text{ or } w \Vdash_1 B$$
$$w \Vdash_0 A \vee B \text{ iff } w \Vdash_0 A \text{ and } w \Vdash_0 B$$
$$w \Vdash_1 A \supset B \text{ iff for all } x \geq w, x \nVdash_1 A \text{ or } x \Vdash_1 B$$
$$w \Vdash_0 A \supset B \text{ iff } w \Vdash_1 A \text{ and } w \Vdash_0 B$$
$$w \Vdash_1 -A \text{ iff } w \Vdash_0 A$$
$$w \Vdash_0 -A \text{ iff } w \Vdash_1 A$$

---

[4]If we want to give up the requirement that there are no statements that are both verified and falsified, we are lead to the system N_4, which utilizes a valuation relation instead of a function, i.e. allows for the assignment of both 1 and 0 to statements.

In N_3 consequence is defined as follows:

$\Gamma \vDash A$ iff in every model and every $w \in W$,

if $w \Vdash_1 B$ for every $B \in \Gamma$, then $w \Vdash_1 A$.

However, this means that in N_3 the property that is transmitted from premises to conlusions is still verifiability. Because we want to transmit non-falsifiability, we define consequence slightly differently:

$\Gamma \vDash A$ iff in every model and every $w \in W$,

if $w \nVdash_0 B$ for every $B \in \Gamma$, then $w \nVdash_0 A$.

Let us call the resulting logic N_{3f}.

The logic N_3 has been studied extensively in (Wansing, 1993, 1998; Odintsov, 2008), and the interested reader can do no better than to turn to these sources for further information. The following table gives an overview over those differences and similarities between N_3 and N_{3f} I want to focus on:

	N_3	N_{3f}
$\vDash A \vee -A$	n	Y
$(A \wedge -A) \vDash B$	Y	n
Double Negation Laws	Y	Y
de Morgan's Laws	Y	Y
$A, A \supset B \vDash B$	Y	n

In N_3, we see Excluded Middle fail, just like it does in intuitionistic logic, and for a very similar reason: We cannot assume, for every statement A, that we can come up with either a verification of A or a verification of $-A$ (that is, at least in the case of N_3, a falsification of A). N_{3f}, on the other hand, validates Excluded Middle. This means that $A \vee -A$ will never be falsified. If you check the clauses, you will find that a falsification of $A \vee -A$ would consist in a verification of A and a falsification of A, something we decided to forbid. Therefore, this validity makes intuitive sense.

The next line tells us that unlike N_3, N_{3f} is paraconsistent, i.e., does not allow for the inference from a contradiction to an arbitrary statement. In terms of the Kripke semantics, if A has no truth value, neither does $-A$, so that it is trivial to construct a counterexample to $A, -A \vDash_{N_{3f}} B$. A single world w and a valuation function that is exhausted by $w \Vdash_0 B$ will do. What this all means on a more philosophical note will be discussed in the next section.

Next in the list are the double negation laws (double negation introduction and elimination) and the de Morgan laws, which are all valid in both N_3 and N_{3f}. In intuitionistic logic, double negation elimination and $-(A \wedge B) \vDash (-A \vee -B)$ are invalid. The thought that there is something non-constructive about these (especially double negation elimination) is deeply ingrained. However, given the new explanation in terms of verifications and falsifications, there is no reason to object to either of them. Double negation elimination fails in intuitionistic logic because of the particular way negation is explained, not because the logic calls for constructive means.

Lastly, modus ponens fails for N_{3f}. This is a, let us say, somewhat worrying feature. We shall deal with it in the next section as well, which will introduce the idea of *incoherence*.

4 Incoherence

When two people are arguing, and both of their assertions will be counted as correct if the audience can not falsify them, then we will expect to see situations in which one party asserts A and the other $-A$, where both of these assertions will be correct. Formally this is reflected in the paraconsistency of the logic: From unfalsified premises A and $\sim A$ it does not follow that any B is unfalsified. Something that *does* follow from such premises, though, is that $(A \wedge \sim A)$ is unfalsified. As we are examining the possibility of equating unfalsifiability and assertibility, we seem to be heading towards allowing people to contradicting themselves, not just their opponents. Should we really be happy to allow people to utter contradictions like that?

Certainly, we should like to ask the participants of our debates to maintain a certain level of coherence in their claims, and part of such coherence is not to utter outright contradictions of the form "A and not A". So, we want to allow the following situation to pass without criticism:

> Alan says: "*A*"
>
> Bertha says: "Not *A*"

What we are unhappy about is this situation:

> Alan says: "*A* and not *A*"

In the last example, Alan was not meeting the coherence constraint and thus we may criticise his assertions.

However, the criticism in this case is different from the kind of criticism that Dummett was talking about when he spoke of falsifiability. Alan did not offend by saying something that we can falsify, because we are in no position to ascertain which of his statements, *A* or *not A*, is false. Incoherence and falsifiability are two different issues, and there are good reasons to keep them apart.

First of all, some may doubt that this kind of incoherence should give rise to criticism in the first place, for example the dialetheists (those who believe in true contradicitons) among us. Second, even if we reject dialetheism, the consequences that the two kinds of flaws have might turn out to be quite different. For example, consider once again a court case. Intentionally saying something that is falsifiable under oath clearly allows the judge to convict the witness of the crime of perjury. On the other hand, simply pointing out that the testimony was incoherent while being unable to prove which part of the testimony was actually false might not be enough to draw that consequence.

Even disregarding the question what constitutes a proof of perjury, the difference between incoherence and falsifiability will affect the further course of the trial. A sly lawyer can shed doubt on a witness's testimony by leading her on to contradict herself. But that doesn't mean that anything in particular she said will from then on count as conclusively falsified: which part of her testimony would that be?

If we *do* want to keep apart the two issues of falsifiability and incoherence, then we should welcome the fact that the logic of falsification does not rule out incoherent statements (unless, of course, they are not only incoherent, but also *falsifiable* as well). Of course, that does not mean that we are not interested in the question whether a set of statements is logically incoherent and thus open for the second kind of criticism.

Before we go on to see what we can do against incoherence, we have to say something about the failure of *modus ponens* as well. This problem is actually analogous to the one involving the unfalsifiability of "A and not A". In the following illustration, let us assume that B is falsified, while nothing is known about A.

First of all, this situation should be unobjectionable:

> Alan says: "A"
>
> Bertha says: "If A, then B."

Both Alan and Bertha make assertions that are unfalsifiable, and thus correct. However, we get into trouble if we speak as Alan does in the next example:

> Alan says: "A. If A, then B."

He really shouldn't be saying both of these things together in view of the fact that B is falsified. We may point out to him that he should refrain from one or the other, but we cannot put our finger on which of the two has to go, as we don't have enough informatioin concerning A. In other words, everything he said is unfalsifiable, but in total his assertions are incoherent and therefore objectionable. The only difference to the asserted contradiction is that what Alan says is not incoherent in itself, but incoherent relative to all states of information in which B is falsified.

5   Getting rid of incoherence

So, what can be done against incoherence? Here is a proposal that is reminiscent to Jaśkowski's discussive logic (Jaśkowski, 1969).

In (Wansing, 1998), a modal extension of Nelson logic is introduced that is to serve as the basis of a non-monotonic reasoning system. The language is extended by a modal operator M. The statement MA is to be read as "it is consistent with what we know at present to assume A".

$w \Vdash_1 A \wedge B$ iff $w \Vdash_1 A$ and $w \Vdash_1 B$
$w \Vdash_0 A \wedge B$ iff $w \Vdash_0 A$ or $w \Vdash_0 B$
$w \Vdash_1 A \vee B$ iff $w \Vdash_1 A$ or $w \Vdash_1 B$
$w \Vdash_0 A \vee B$ iff $w \Vdash_0 A$ and $w \Vdash_0 B$
$w \Vdash_1 A \supset B$ iff for all $x \geq w$, $x \nVdash_1 A$ or $x \Vdash_1 B$
$w \Vdash_0 A \supset B$ iff $w \Vdash_1 A$ and $w \Vdash_0 B$
$w \Vdash_1 -A$ iff $w \Vdash_0 A$
$w \Vdash_0 -A$ iff $w \Vdash_1 A$
$w \Vdash_1 MA$ iff there is an $x \geq w$, $x \Vdash_1 A$
$w \Vdash_0 MA$ iff for all $x \geq w$, $x \Vdash_0 A$

Even though we still postulate persistence of both 0 and 1 for atomic statements in these models, it is easy to come up with models that show that this does not hold for arbitrary formulas any more. However, for formulas not containing the modal operator, the heredities still hold true.

Now, how might extending our language to one with a modal consistency operator help with our problem of incoherent speech?

The most straightforward proposal is this: We decide whether a set of statements $A_1, \ldots, A_n$ can be coherently asserted by checking whether their conjunction is possible, that is whether our current state of information supports $M(A_1 \wedge \ldots \wedge A_n)$.

Simple Modal Coherence A set of statements $A_1, \ldots, A_n$ is coherently assertible at a world w iff $w \Vdash_1 M(A_1 \wedge \ldots \wedge A_n)$

This solution indeed does away with assertions of contradictions (generally incoherent assertions), as $w \nVdash_1 M(A \wedge -A)$ for all worlds w. Assertions that are only incoherent in view of the information at a given world can also be discounted. It can be easily checked that if a state $w \Vdash_0 B$, then $w \nVdash_1 M(A \wedge A \supset B)$.

However that might be, the modal strategy throws out too much unobjectional material along with the things we indeed want to get rid of. From all we've seen up to now by way of motivational arguements for unfalsifiability as an assertion norm, there is no reason at all to discount assertions that are unfalsified at a world, even if we have conclusive proof that no verification for the asserted statement will ever be obtained, as long as that proof of unobtainability involves only information about our cognitive limitations. Let's consider the following statement, uttered in our times:

> Julius Caesar had an even number of hairs on his head when he was stabbed to death.

It is very unlikely that we should in the future find historical records that deal with this issue or devise a method that could deliver good evidence for or against this statement. So, we may consider it verified that we will never verify the statement above. The Simple Modal Coherence criterion will thus rule this statement out as incoherent. Nonetheless, and importantly differently from the case of an asserted contradiction, this should not give us grounds to dismiss the statement.

However, what the modal statement we want to use as a filter says is that a statement is coherent iff it is possible that it will be verified at some point. This is false in the described scenario. Thus, the claim will be stigmatized as incoherent under the strictures of Simple Modal Coherence, something we should not want.

Here is a proposal to get around this problem.[5] It involves adding yet more modal vocabulary to our language to make reference to the past possible: Then we will be able to say that a (set of) statement(s) is coherent if, at some point in the past, there had been a possibilty that it would become verified later on. That is, we take the claim about Caesar's hair to be coherent because we can discern a point in the past, say the morning of his assassination, where the verification of it was a possibility (even if it would have been unlikely even back then that someone should have actually counted the number of hairs on his skull).

The addition to our vocabulary is straightforward: A modal operator P that is defined as follows:

$$w \Vdash_1 PA \text{ iff there is an } x \leq w, x \Vdash_1 A$$
$$w \Vdash_0 PA \text{ iff for all } x \leq w, x \Vdash_0 A$$

---

[5]Thanks to Graham Priest for suggesting this fix.

Now we can spell out Priest's proposal for coherence:

Elaborate Modal Coherence A set of statements A_1, ..., A_n is coherently assertible at a world w iff $w \Vdash_1 PM(A_1 \wedge \ldots \wedge A_n)$

The Caesar problem is fixed with this extension. But what about the following statement?

> There are large quantities of gold on Mars, but we will never find them or any trace of them.

This is not verifiable, because to verify it we'd have to verify both conjuncts. To verify the first, however, is to falsify the second. The problem is that this an inherent feature of this sentence: There was never a time in which it was still a possibility that it would be verified later on. So on the new elaborate modal scheme, this sentence is incoherent.

Should it be, though? We might be entering an area where it would be possible to dig in our heels and applaud the fact that the new criterion proclaims the statement incoherent. There *is* a slightly weird Moorean quality about this sentence, especially in view of the anti-realistic assumptions that Dummett's program is built on.

What if we want to allow such statements to stand as coherent? No matter how sophisticated with the modalities in our attempt to capture coherence, it is clear that we will never get to the point of allowing these kinds of sentences in.

Here's last proposal that we can turn to when the modal approach gives out. It is, quite simply, to check the set of assertions we want to assess for classical consistency with the information at hand. But this seems to be the point where an anti-realistic story that will motivate such a move will be very hard to give. If classical logic can not be given a basis that can satisfy anti-realistic demands, then to appeal to it to account for our notion of coherence would not seem to be a legitimate move.

References

Dummett, M. (1993). *The seas of language.* Oxford: Clarendon Press.

Jaśkowski, S. (1969). Propositional calculus for contradictory deductive systems. *Studia Logica, 24*(1), 143–157.

López-Escobar, E. (1972). Refutability and elementary number theory. *Indagationes Mathematicae, 34*, 362–374.

Odintsov, S. (2008). *Constructive negations and paraconsistency.* Springer Verlag.

Wansing, H. (1993). *The logic of information structures.* Berlin: Springer.

Wansing, H. (1998). *Displaying modal logic.* Kluwer Academic Publishers.

Andreas Pietz
LOGOS Group
Department of Logic
Faculty of Philosophy
University of Barcelona
C. Montalegre, 6–8
Barcelona 08001
e-mail: andreas.pietz@gmail.com

Revenge of the Indexical Liar

Martin Pleitz*

Abstract

In this paper I show that the indexical variant of the Liar paradox cannot be solved by recourse to the need to give local truth conditions to indexical sentences (thus I give a negative answer to a hopeful question I asked in my contribution to *The Logica yearbook 2009*). After describing the particular case of the indexical Liar sentence in some detail (§2), I develop a general theory of localized truth, characterizing the four different notions of truth that are needed in an indexicalist setting (§3). On that basis I distinguish three different readings of the indexical Liar sentence and show that, while two of these readings are harmless (§4), the third one is paradoxical (§5). After generalizing this paradoxicality result (§6), I conclude by contrasting it with other revenge problems that confront indexicalist and contextualist approaches to the Liar paradox (§7).

1   Awkward for the looker-on

The Liar paradox involves a *Liar sentence*, i.e., a sentence that says of itself that it is not true. Given the Tarskian truth schema, which explicates a highly plausible principle about truth, and classical logic, the claim that a Liar sentence exists entails a contradiction.

Suppose, for example, that the following somewhat circular[1] definition of a Liar sentence is in order:

---

*I would like to thank Johannes Korbmacher, Peter Rohs, and Ansgar Seide for helpful questions and comments. This text is part of a project that is dedicated to the memory of Rosemarie Rheinwald.

[1] I hold (and will try to show in other work) that this circularity is much more problematic than it looks to many a contemporary logician, and that herein lies the key to solving the Liar paradox. But in the present paper I am concerned only with the indexical Liar sentence 'This sentence is not true' (cf. §2). The standard Liar

(1) $\Lambda =_{\mathrm{def}}$ 'Λ is not true'

Then Λ quite obviously means that Λ is not true. Hence, Λ is a Liar sentence.

From (1), the logic of identity (in particular, the indiscernability of identicals) allows to derive:[2]

(2) Λ is true $\Leftrightarrow$ 'Λ is not true' is true

It is a highly plausible principle about truth that a sentence is true if and only if what it says is the case. This principle can be explicated as the Tarskian truth schema, which amounts to an assertion of every sentence that is obtained from the sentential form 's is true if and only if p' by substituting for 's' a singular term that refers to some sentence and substituting for 'p' a translation of that sentence (Tarski, 1956, pp. 187f., notation altered). Now, ''Λ is not true'' is a singular term that refers to Λ and Λ can, in view of what it says, be translated homophonically as 'Λ is not true'. Therefore the Tarskian truth schema entails for Λ:

(3) 'Λ is not true' is true $\Leftrightarrow$ Λ is not true

Classical propositional logic allows to derive a contradiction from (2) and (3):

(4) Λ is true $\Leftrightarrow$ Λ is not true

To see how deeply troubling this contradictory result is, note that in this paradoxical reasoning the Liar sentence Λ is not used, but only mentioned.[3] As a consequence, we can see that not only any theory that includes a Liar sentence like Λ as a theorem must be an inconsistent theory and any language that includes a Liar sentence must be an inconsistent language, but even any language that merely

---

reasoning, which is illustrated here by the circularly defined Λ, provides nothing more than a contrast for the indexical variant of the Liar paradox.

   [2] The sign '$\Leftrightarrow$' abbreviates the English locution 'if and only if'; '$\Leftarrow$' abbreviates 'if ... then'.

   [3] A homophonic translation of Λ is indeed used in the right-hand sides of (3) and (4). But nothing depends on a Liar sentence being a sentence of the language we reason in—it suffices to know that it is a sentence that attributes untruth to itself.

allows to *talk about* a Liar sentence must be inconsistent. The *deeply troubling fact* about the Liar paradox is that the contradiction is not confined to the object language that the Liar sentence belongs to, but arises in any meta-language we use to reason about it.

Arthur Prior makes a similar observation:

> "Some paradoxical statements are, on the face of it, awkward for the propounder only, while some are also awkward for the looker-on. The Eubulidean version of the Liar paradox is of the second sort—if a man says 'What I am now saying is false', not only he himself but we who look on seem forced to say contradictory things (that his statement must be true because even if it were false it would be true, *and* that it must be false because even if it were true it would be false)." (Prior, 1961, p. 16)

But maybe Prior is too pessimistic—at least with respect to the instance of a Liar sentence he gives. As 'what I am now saying' is an *indexical* singular term, there may be some hope that at least in this case the awkwardness is confined to the propounder. Might not the contradiction that arises from an indexical Liar sentence be nothing more than *a harmless singularity in logical space*, i.e., a problem that concerns only that very sentence but does not spread beyond it?

In this paper, I will give a negative answer: Even indexical Liar sentences are not only paradoxical but what I call deeply troubling. To show this, I will describe the particular case of the indexical Liar sentence in some detail (§2) and develop a general theory of localized truth that allows to deal with indexical sentences (§3). On that basis I will distinguish three different readings of the indexical Liar sentence and show that, while two of these readings are harmless (§4), the third one is paradoxical (§5). After generalizing this paradoxicality result (§6), I will contrast my results with other revenge problems that confront indexicalist and contextualist approaches to the Liar paradox (§7).

2 The indexical liar sentence and its local truth condition

The Eubulidean singular term 'what I am saying now' combines personal indexicality ('I'), temporal indexicality ('now') and the notion

of saying something in order to refer to the linguistic expression it is
a part of. This turns out to be an avoidable detour because the same
kind of self-reference can be achieved in a direct way by a regimented
use of the locution 'this sentence'. Regimentation is necessary because
'this sentence' is ordinarily used to refer to *another* sentence.[4] But
for the context of the present investigation we can just *stipulate* that
every occurrence of the term 'this sentence' refers to the sentence it is
used in. This gives us a more perspicuous Liar sentence (call it 'λ'):

(λ) 'This sentence is not true.'

Can λ be the starting point of a derivation of a contradiction that
is similar to the above derivation that starts from Λ? Prior seems to
have thought so in his reasoning about the truth or falsity of 'What I
am now saying is false', thus in effect assimilating indexical to context-
free reasoning. When Prior wrote about the Eubulidean statement in
1961, ignoring the particularities of indexicality was still quite com-
monplace. But in 1967, Donald Davidson pointed out that indexicals
"cannot be eliminated from natural language without loss or radical change" (Davidson, 1967, pp. 318f.). For sentences containing
indexicals, he substituted the Tarskian truth schema by local truth
conditions like the following:

> "'I am tired' is true as (potentially) spoken by p at t if
> and only if p is tired at t." (Davidson, 1967, pp. 319f.)

Should we give similar local truth conditions to the indexical Liar
sentence? Will that dispel its paradoxicality? In my contribution to
The Logica Yearbook 2009, I argued that the first question must be
answered 'yes' (Pleitz, 2010, pp. 197–203) and expressed the hope
that an affirmative answer to the second question might also be
found (Pleitz, 2010, pp. 203–205). Today, although I stand with my
first answer, my associated hopes of its being the basis for a solution to
the Liar paradox are thoroughly dashed. But before I show how local
truth conditions allow for an alternative derivation of a contradiction

---

[4]In their work on circular propositions, Jon Barwise and John Etchemendy
make a similar regimentation, pointing out that "'this proposition' can also
be used demonstratively, to refer to some other proposition immediately at
hand" (Barwise & Etchemendy, 1987, p. 16).

(§3, §5), I will briefly recount why we need local truth conditions for the indexical Liar sentence and how this gives reason for hope.

An expression is *indexical* if and only if its extension depends in a systematic way on its position of use (despite stability of meaning). The word 'now' is *temporally indexical* because it refers to the moment of its use and the word 'I' is *personally indexical* because it refers to the person who uses it. As the term 'this sentence' refers to whatever sentence it is a part of, it should be understood as *sententially indexical* and any sentence it occurs in should be construed as its position of use, analogous to a moment or a speaker. Hence, sentences that contain the term 'this sentence' should be given local truth conditions, based on the notion of *truth at a sentence*. In my Logica 2009 contribution, I gave the following schema for the local truth conditions of atomic sentences with the term 'this sentence' in subject position (cf. Pleitz, 2010, p. 199):[5]

(L) 'This sentence is Q' is true at sentence $s \Leftrightarrow s$ is Q

Now, to generate a local truth condition for λ, we need only substitute 'not true' for 'Q':

(L_λ) 'This sentence is not true' is true at sentence $s \Leftrightarrow s$ is not true

By instantiating the variable 's' as λ (and substituting 'λ' for the quotation of λ) we get:

(B) λ is true at $\lambda \Leftrightarrow \lambda$ is not true

As *local truth* (expressed by the dyadic predicate 's is true at t') is not the same notion as *truth simpliciter* (expressed by the monadic predicate 's is true'), the biconditional (B) is not contradictory. So, may there not be hope of solving the Liar paradox, at least in its indexical variant? I quote my past assessment:

"Of course, the mere fact that biconditional (B) is not itself contradictory does not show conclusively that there

---

[5]While 'Q' is a *schematic letter* to be substituted by some predicate applicable to sentences, 's' is a *variable* that ranges over sentences and is tacitly understood to be bound by a universal quantifier (perhaps restricted to sentences that contain the term 'this sentence').

is no way of deriving a contradiction from the localized
truth condition of the indexical Liar sentence (L$_\lambda$). But
the fact that the standard derivation does not work any
longer gives reason to hope that the indexicalist approach
to the Liar paradox might lead to its solution." (Pleitz,
2010, p. 204)

A first sign of trouble shows up when we compare the schema (L) to
Davidson's example for a local truth condition. While Davidson translates the object language predicate '*p* is tired' into the meta-language
as '*p* is tired at *t*' and thus moves from a monadic to a dyadic predicate, in (L) the object language predicate '*x* is *Q*' is translated as the
meta-language predicate '*x* is *Q*', i.e., homophonically. In the temporal case, surely *some* predicates can be translated homophonically
from an indexical object language to a context-free meta-language—to
wit, '*x* is human', which is temporally context-free to begin with. But
if it is indexical, an *n*-adic predicate can only be adequately translated
into a context-free language by a corresponding $n + 1$-adic predicate
that has an extra place for the relevant position of use.

To give the general form of the local truth condition of an atomic
sentence in a setting where truth can be relative to some parameter,
we have to be aware that *both* the reference of a singular term *and*
the extension of a predicate can be relative to that parameter. An
atomic sentence is true relative to *x* if and only if the singular term's
reference relative to *x* is an element of the extension the predicate has
relative to *x*. More formally:[6]

$$\ulcorner \alpha \text{ is } \Phi \urcorner \text{ is true relative to } x \Leftrightarrow [\![\alpha]\!]_x \in [\![\Phi]\!]_x$$

Therefore local truth conditions like (L) are adequate only when
a context-free predicate is substituted for '*Q*'. — So, what about the
predicate '*s* is not true'? Is there a good sense in which a monadic
truth predicate can be construed as *indexical*?

3   A general theory of local truth

To develop a general theory of truth for an indexical setting, I will
now turn for inspiration to the well-understood case of temporal in-

---

[6]Here, 'α' and 'Φ' are meta-variable that range over singular terms and predi-
cates and $\ulcorner [\![\varphi]\!]_x \urcorner$ designates the extension that the expression φ has relative to x.

dexicality. Compare the following ascriptions of truth to a sentence:

(τ_1) ''$2 + 2 = 4$' is true.'

(τ_2) ''It is snowing' is true on January 1, 2011.'

(τ_3) ''It is snowing or it is not snowing' is true.'

(τ_4) ''It is snowing' is true.'

There are four different notions of truth in play here: τ_1 ascribes *atemporal truth*, τ_2 ascribes *momentary truth*, τ_3 ascribes *omnitemporal truth* and τ_4 ascribes *tensed truth*. These notions can be characterized in the following way. The notions of atemporal truth and momentary truth are primitive; they are mutually exclusive and exhaustive in the following sense:

> No sentence has both an atemporal truth value and a momentary truth value. Every sentence either has an atemporal truth value or a momentary truth value.[7]

Omnitemporal and tensed truth can be defined by recourse to momentary truth:

A sentence is *omnitemporally true* $\Leftrightarrow$

it is true at every moment[8]

$\ulcorner \alpha$ is TENSEDLY TRUE$\urcorner$ is true at a moment t $\Leftrightarrow$

the sentence that α refers to at t is true at t

While the notions of atemporal truth, momentary truth, and omnitemporal truth are context-free, the notion of tensed truth is indexical. As tensed truth must be expressed by an indexical predicate, it cannot be defined directly in context-free terms alone but only by semantic ascent. Therefore the framework for the description of the four

---

[7]To allow for non-bivalent languages, we have to construe a gap as a further truth value. In the present setting we will of course have to be careful not to confuse atemporal gaps and omnitemporal gaps.

[8]Note that this definition of omnitemporality is a case of supervaluation. — It is natural to add to it that a sentence is *omnitemporally false* if and only if it is false at every moment. Then omnitemporal truth will in general be gappy even if truth at a moment is not. — Atemporal truth will be gappy for a similar reason.

notions of truth should be comprised of a context-free meta-language (which can express atemporal truth, momentary truth, and omnitemporal truth) and an indexical object language (that can also express indexical truth). The context-free predicates can be translated homophonically into the indexical object language:[9]

> $\ulcorner \alpha$ is TRUE AT MOMENT $t \urcorner$ is true at moment $s \Leftrightarrow$
>> the sentence that α refers to at s is true at moment t

> $\ulcorner \alpha$ is ATEMPORALLY TRUE $\urcorner$ is true at moment $t \Leftrightarrow$
>> the sentence that α refers to at t is atemporally true

> $\ulcorner \alpha$ is OMNITEMPORALLY TRUE $\urcorner$ is true at moment $t \Leftrightarrow$
>> the sentence that α refers to at t is omnitemporally true

As temporal indexicality is ubiquitous, the notions of momentary truth, atemporal truth, omnitemporal truth, and tensed truth have many applications in ordinary discourse. Therefore it should be easy to check that the present account of these notions and their interrelations is an adequate explication.

Let us now move on to less well-known dimensions of indexicality. The first step is to generalize—from atemporal truth to *apositional truth*, from moments of time to *positions in general*, from *omnitemporal truth* to omnipositional truth, and from tensed truth to *indexical truth*. By analogy to temporal indexicality, we obtain the following four general notions:

Local Truth Local truth is a primitive notion; we say: $\ulcorner$Sentence s is *locally true* at the position of use $p. \urcorner$

Apositional Truth Apositional truth is a primitive notion; we say: $\ulcorner$Sentence s is *apositionally true.* $\urcorner$

Exclusiveness and Exhaustiveness No sentence has both an *atemporal truth value* and a *momentary truth value*. Every sentence either has an *atemporal truth value* or a *momentary truth value*.[10]

---

[9]To distinguish them from truth predicates of the context-free meta-language, I write every truth predicate of the indexical object language in SMALL CAPITALS.

[10]In the case of non-bivalent languages, we have to allow for apositional gaps and local gaps which are construed as further truth values.

Omnipositional Truth A sentence is *omnipositionally true* $\Leftrightarrow$ it is locally true at every position

Indexical Truth A sentence of the form $\ulcorner\alpha$ is INDEXICALLY TRUE$\urcorner$ is locally true at a position $p \Leftrightarrow$ the sentence α refers to at p is locally true at p

In a somewhat abridged notation,[11] and including the homophonical translation of the predicates expressing local, omnipositional and apositional truth into the indexical object language, we get:

(X) φ has an apositional truth value $\Leftrightarrow$ φ does not have a local truth value

(O) φ is true$_O$ $\Leftrightarrow$ for every position p, φ is true$_L$ at p

(L_L) $\ulcorner\alpha$ is TRUE$_L$ AT $p\urcorner$ is true$_L$ at $q \Leftrightarrow [\![\alpha]\!]_q$ is true$_L$ at p

(L_A) $\ulcorner\alpha$ is TRUE$_A$ $\urcorner$ is true$_L$ at $p \Leftrightarrow [\![\alpha]\!]_p$ is true$_A$

(L_O) $\ulcorner\alpha$ is TRUE$_O$$\urcorner$ is true$_L$ at $p \Leftrightarrow [\![\alpha]\!]_p$ is true$_O$

(L_I) $\ulcorner\alpha$ is TRUE$_I$$\urcorner$ is true$_L$ at $p \Leftrightarrow [\![\alpha]\!]_p$ is true$_L$ at p

That is all.

Note that, as in the temporal case, the three context-free notions of local, apositional and omnipositional truth can be expressed both in the context-free meta-language and the indexical object language, but the notion of indexical truth can be characterized only by semantic ascent.

To complete our way from temporal to sentential indexicality, we have to take a second step, construing sentences as positions of use for sentences that contain the sententially indexical singular term 'this sentence'. What we get, besides the localized notion of *truth at a sentence* familiar from §2, are the notions of *asentential truth* (characterized by (X) and (L_A)), *omnisentential truth* (defined by (O) and (L_O)) and *sententially indexical truth* (defined by (L_I)).

---

[11] Here, 'true$_L$' expresses local truth, 'true$_A$' expresses apositional truth, 'true$_O$' expresses omnipositional truth and 'TRUE$_I$' expresses indexical truth. Again, SMALL CAPITALS are used to distinguish object language from meta-language truth predicates. 'p' and 'q' range over positions, 'φ' ranges over sentences, 'α' ranges over object language singular terms that refer to sentences and $\ulcorner[\![\alpha]\!]_p\urcorner$ designates the object that α refers to at p.

temporal indexicality:	indexicality in general:	sentential indexicality:	notation:
truth at a moment	local truth (relation)	truth at a sentence	'true$_L$'
omnitemporal truth	omnipositional truth	omnisentential truth	'true$_O$', 'TRUE$_O$'
atemporal truth	apositional truth	asentential truth	'true$_A$', 'TRUE$_A$'
tensed truth	indexical truth	sentential truth	'TRUE$_I$'

Table 1: the way from temporal to sentential indexicality

4   Three readings of the indexical Liar sentence

Now we are in a position to disambiguate the indexical Liar sentence λ, 'This sentence is not true'. While the predicate that expresses truth at a sentence is dyadic ('s is true$_L$ at t'), the other three notions of truth are expressed by monadic predicates ('s is true$_A$', 's is true$_O$', and 's is TRUE$_I$'). Hence there are three possible meanings for the monadic truth predicate in λ, which lead to three readings of the indexical Liar sentence:

(λ_A)   'This sentence is not TRUE$_A$.'

(λ_O)   'This sentence is not TRUE$_O$.'

(λ_I)   'This sentence is not TRUE$_I$.'

In order to make the truth-conditions for truth predicates given in §3 applicable to these Liar sentences, we need the following standard condition for negation:[12]

(N) $\ulcorner$Not $\varphi\urcorner$ is true$_L$ at s $\Leftrightarrow$ φ is not true$_L$ at s

---

[12]The condition (N) is similar to the clauses for negation in the Tarskian recursive definition of truth, in possible worlds semantics, and in two-dimensional semantics.

Now we can show that the Asentential Indexical Liar sentence λ_A and the Omnisentential Indexical Liar sentence λ_O are harmless. (The Fully Indexical Liar sentence[13] λ_I will be the topic of §5.)

With regard to the Asentential Indexical Liar λ_A we can reason as follows. As λ_A contains the indexical term 'this sentence', it must have local truth conditions (cf. §2). By (X), λ_A has no asentential truth value. Hence, λ_A is not true$_A$. By (L$_A$) and (N), λ_A is true$_L$ at λ_A. Therefore:

(C$_A$) λ_A is not true$_A$ and λ_A is true$_L$ at λ_A.

As we have used all principles that connect the notions of local and asentential truth ((X) and (L$_A$)), it is highly unlikely that there is any way in which a contradiction can be derived from the existence of the sentence λ_A.

With regard to the Omnisentential Indexical Liar λ_O, our reasoning differs only slightly. As λ_O contains the indexical term 'this sentence' it must have local truth conditions. Observe that λ_O is true$_L$ at the omnisentential truth 'This sentence is not TRUE$_O$ or it is not the case that this sentence is not TRUE$_O$' and that λ_O is not true$_L$ at the omnisentential untruth 'This sentence is not TRUE$_O$ and it is not the case that this sentence is not TRUE$_O$'. Hence, by (O), λ_O is not true$_O$, and so by (L$_O$) and (N), λ_O is true$_L$ at λ_O.

(C$_O$) λ_O is not true$_O$ and λ_O is true$_L$ at λ_O.

As we have used all principles that connect the notions of local and omnisentential truth ((O) and (L$_O$)), it is highly unlikely that there is any way in which a contradiction can be derived from the existence of the sentence λ_O.

How are these results about the harmlessness of λ_A and λ_O related to the hopes for a solution of the indexical variant of the Liar paradox? Note first that both (C$_A$) and (C$_O$) entail a biconditional of a form familiar from §2,

(B) λ is true at $\lambda \Leftrightarrow \lambda$ is not true,

---

[13]I call λ_I '*fully* indexical' because it combines a sententially indexical singular term ('this sentence') and a sententially indexical predicate ('*s* is not TRUE$_I$').

with 'λ_A' or 'λ_O' substituted for each occurrence of 'λ' in (B). Thus the harmlessness results reached here can be seen as a partial[14] fulfillment of the hopes described in §2. The agreement between (C_A) and (C_O) on the one hand and (B) on the other is easily explained by a similarity on the level of local truth conditions. (L_O) and (L_A) are similar to the schematic local truth condition (L_λ) in §2 insofar as each one of them translates a truth predicate homophonically. But by we now have more information, provided by the general theory of local truth developed in §3, and in particular by (X) and (O). These conditions exclude one of the cases left open by a biconditional of the form (B) and thus lead us to the conjunctions (C_A) and (C_O).

5   The paradoxicality of the Fully Indexical Liar sentence

The Fully Indexical Liar sentence λ_I, 'This sentence is not TRUE$_I$', is paradoxical. This can be shown in the following way. Recall the local truth condition for a sentence that ascribes sententially indexical truth and the local truth condition for negation:

(L$_I$) ⌜α is TRUE$_I$⌝ is true$_L$ at s ⇔ $[\![\alpha]\!]_s$ is true$_L$ at s

(N) ⌜Not φ⌝ is true$_L$ at s ⇔ φ is not true$_L$ at s

Now instantiate 'φ' in (N) as 'This sentence is TRUE$_I$':

(1) 'This sentence is not TRUE$_I$' is true$_L$ at s ⇔ 'This sentence is TRUE$_I$' is not true$_L$ at s

From (L$_I$) we get by instantiating 'α' as 'this sentence':

(2) 'This sentence is TRUE$_I$' is true$_L$ at s ⇔ $[\![$'this sentence'$]\!]_s$ is true$_L$ at s

In view of the semantics of the sententially indexical locution 'this sentence', we know that $[\![$'this sentence'$]\!]_s = s$. Therefore (2) entails:

(3) 'This sentence is TRUE$_I$' is true$_L$ at s ⇔ s is true$_L$ at s

---

[14]The fulfillment is only partial because it concerns only two of three readings of the indexical Liar sentence.

By propositional logic we can move on to:

(4) 'This sentence is TRUE$_I$' is not true$_L$ at $s \Leftrightarrow s$ is not true$_L$ at s

(1) and (4) together entail:

(5) 'This sentence is not TRUE$_I$' is true$_L$ at $s \Leftrightarrow s$ is not true$_L$ at s

Finally, we only have to substitute 'λ_I' for the quotation of λ_I and to instantiate the variable 's' as λ_I:

(6) λ_I is true$_L$ at $\lambda_I \Leftrightarrow \lambda_I$ is not true$_L$ at λ_I

The above derivation of the contradiction (6) does not only show that λ_I is paradoxical but also that the paradoxicality of λ_I is deeply troubling in the sense explained in §1. As both the singular term and the predicate in λ_I are indexical, there might have been more reason here than anywhere else to hope for the paradoxicality to concern only the sentence that is its origin. But λ_I is paradoxical not only from the perspective of λ_I. Neither the indexical singular term 'this sentence' nor the indexical predicate 's is TRUE$_I$' is used at any point in the reasoning about λ_I—the entire derivation is context-free. From the fact that, to produce paradox, it suffices to *mention* the problematic indexical locutions in a context-free meta-language we can see that λ_I is as troubling as any other Liar sentence (e.g., Λ in §1). It is not only "awkward for the propounder", i.e., for itself, but also "awkward for the looker-on",[15] i.e., for anyone with the resources to reason about an indexical language and in particular about indexical truth.

And so my earlier hopes that the need for local truth conditions will dispel the paradoxicality of the indexical Liar sentence (cf. §2) are finally and completely dashed. When the indexical Liar sentence λ is understood as attributing to itself not the asentential or the omnisentential, but the *sententially indexical* notion of untruth, then the clause (L$_I$) of the theory of local truth (cf. §3) leads straight to paradox.

---

[15](Prior, 1961, p. 16, cf. §1).

6   A localized variant of the Tarskian truth schema

What has happened? Note that in the reasoning about the Fully
Indexical Liar sentence in §5 no instance of the Tarskian truth schema
is employed. From the clauses governing the semantics of negation,
of the locution 'this sentence', and of the sententially indexical truth
predicate (i.e., from (N), $\lceil [\![\text{'this sentence'}]\!]_s = s \rceil$, and ($L_I$)), we first
deduce a local truth condition for λ_I (i.e., (5)), and from there we
can move on to a contradiction in much the same way as we can
from an instance of the Tarskian truth schema in the case of a non-
indexical Liar sentence (cf. the step from (3) to (4) in §1). So, in
the reasoning about the Fully Indexical Liar sentence, a local truth
condition occupies the role of the Tarskian truth schema. It is highly
plausible that this fact can be generalized: The semantic theory of
indexicality provides no way out of the Liar paradox.

And why should it? In the context of his plea for giving local
truth conditions for indexical sentences, Davidson notes that if we
"view truth as a relation between a sentence, a person, and a time",
then "ordinary logic as now read applies as usual, but only to sets of
sentences relativized to the same speaker and time" (Davidson, 1967,
p. 319). And, in a way, the Tarskian truth schema also applies as
usual—but only in a way relativized to positions of use. When we
combine the local truth condition (L_I) that gives the semantics of the
indexical notion of truth with the usual localized semantics for propositional connectives, we can derive *a localized variant of the Tarskian
truth schema*: For every position x, if a sentence is substituted for 'p'
and a term that refers at x to that sentence is substituted for 's', then

's is TRUE$_I$ if and only if p' is true$_L$ at x.[16]

Or, more formally:

$$\forall p([\![\alpha]\!]_p = \varphi \implies \lceil \alpha \text{ is TRUE}_I \Leftrightarrow \varphi \rceil \text{ is true}_L \text{ at } p)$$

These considerations about localizing the Tarskian truth schema
are in no way bound to the (arguably somewhat strange) *sentential*

---

[16]Thus Tarski was right, after all, whenever he said "'It is snowing' is true if and
only if it is snowing" (Tarski, 1956, p. 156)—although, in view of the context-free
framework he without doubt aimed at, this example of an ascription of *indexical*
truth remains puzzling.

dimension of indexicality. We easily arrive at a result that is considerably more general than the paradoxicality of λ_I:

(P) Any sentence of the form $\ulcorner\alpha$ is not TRUE$\urcorner$ is paradoxical if

    (i) the singular term α is P-indexical, i.e. its reference depends on its position of use in the dimension of indexicality P,

    (ii) the predicate 's is TRUE' expresses the P-indexical notion of truth and

    (iii) there is a position $q \in$ P such that $[\![\alpha]\!]_q = \ulcorner\alpha$ is not TRUE$\urcorner$.[17]

This can be shown by a variant of the derivation (1) through (6) in §5[18]—or we can reason more briefly:

(1) $\forall p(\ulcorner\alpha$ is TRUE$_I\urcorner$ is true$_L$ at $p \Leftrightarrow [\![\alpha]\!]_p$ is true$_L$ at p)

                                             (i), (ii), (L$_I$)

(2) $\forall p(\ulcorner\alpha$ is not TRUE$_I\urcorner$ is true$_L$ at $p \Leftrightarrow [\![\alpha]\!]_p$ is not true$_L$ at p)

                                                (1), (N)

(3) $\exists q([\![\alpha]\!]_q = \ulcorner\alpha$ is not TRUE$_I\urcorner$)

                                                    (iii)

(4) $\exists q(\ulcorner\alpha$ is not TRUE$_I\urcorner$ is true$_L$ at $q \Leftrightarrow \ulcorner\alpha$ is not TRUE$_I\urcorner$ is not true$_L$ at q)

                                                   (2), (3)

---

[17]Further presuppositions are the rules of classical logic (which are localized in a straightforward way) and what many call the "naïve" conception of truth (which here is explicated not by the Tarskian truth schema but by the local truth condition of an indexical truth predicate; cf. §3). Hence, the only hope for a solution of the Liar paradox that saves both classical logic and the naïve conception of truth lies in trying to show that clause (iii) of (P) cannot be satisfied. I surmise that reasons for this can be found in the (local) *circularity* of the purported Liar sentence $\ulcorner\alpha$ is not TRUE$_I\urcorner$ that is expressed in (iii) by '$[\![\alpha]\!]_q = \ulcorner\alpha$ is not TRUE$_I\urcorner$'.

[18]We only have to understand the clauses (N) and (L$_I$) so that they concern a P-indexical language, to let the variable 's' range over P, to substitute the meta-variable 'α' for each occurrence of the particular locution 'this sentence' and accordingly to change all quotation marks into Quine corners.

7   Indexical revenge

What we have here is a case of the *revenge* of the Liar[19]—in an indexical setting. In general, the revenge phenomenon is that after an initially plausible attempt at solving the Liar paradox there crops up another variant that is closely related to the very method employed in the proposed solution. In an indexical setting, the proposed solution is to insist on a local truth condition, which blocks the standard derivation of a contradiction. The *indexical revenge* consists in the paradoxicality of any indexical Liar sentence that fulfills the conditions of (P), of which the paradoxicality of the Fully Indexical Liar sentence is a particular case.

Keith Simmons distinguishes between *direct revenge* and *second order revenge*:

> "In the simplest manifestation of revenge—call it *direct revenge*—the pathological sentence or expression re-emerges intact from the attempt to treat it." (Simmons, 2007, p. 345) "But revenge can take another form—call it *second order revenge*. Often a solution to a paradox will introduce new, perhaps technical notions—for example, gaps [...], levels of a hierarchy, groundedness, determinate truth, stability, context. Second-order revenge takes these new notions, and constructs new paradoxes for old." (Simmons, 2007, p. 347)

Given the present state of work on the Liar paradox, direct revenge must be seen as an even graver problem than second order revenge. While direct revenge is quite rare, many (if not all) proposed solutions presently under discussion are saddled with second order revenge: They cannot express notions that are crucial to the solution on pain of revenge phenomena. It therefore attests to the severity of the problem posed by the indexical revenge phenomenon described by (P) that it must be construed not as second order revenge, but as direct revenge.

Many contemporary approaches to the Liar are based on the indexicality or context-dependence of truth. Important examples are (Parsons, 1974), (Burge, 1979), (Barwise & Etchemendy, 1987),

---

[19]Cf. (Beall, 2007).

(Simmons, 1993), and (Glanzberg, 2003). They have often been confronted with revenge sentences like the following (cf., e.g. Gauker, 2006, pp. 412f.):

(ρ_1) 'ρ_1 is not TRUE$_L$ in any context.'

(ρ_2) 'ρ_2 is not TRUE$_L$ in this context.'

As explicit reference to contexts need not be among the resources of a paradox-free object language, ρ_1 and ρ_2 are cases of second order revenge. The indexical Liar sentences that satisfy the conditions of (P), on the other hand, employ only an indexical truth predicate, which will doubtlessly be needed in any indexical language that is to allow ascriptions of truth.

Compare ρ_2, which explicitly deals with contexts and with context-relative truth, to

(ρ_3) 'ρ_3 is not TRUE$_I$.'

According to (P), ρ_3 is paradoxical if there is a context where 'ρ_3' refers to ρ_3. But in the sentence ρ_3, none of the apparatus needed to describe context-sensitivity is used. This shows that the Fully Indexical Liar sentence λ_I and its ilk are much more direct than the revenge sentence proposed by Christopher Gauker in his penetrating critique of contextualist approaches to the Liar paradox (Gauker, 2006, pp. 412f.), which is similar to ρ_2.

Because of its generality and directness, the phenomenon of indexical revenge described in this paper must be seen as a grave new addition to the revenge problems that already confront indexicalist and contextualist approaches to the Liar paradox.

References

Barwise, J., & Etchemendy, J. (1987). *The Liar.* New York and Oxford: Oxford University Press.

Beall, J. (2007). *Revenge of the Liar. New essays on the paradox.* Oxford: Oxford University Press.

Burge, T. (1979). Semantical paradox. *The Journal of Philosophy, 76,* 169–198.

Davidson, D. (1967). Truth and meaning. *Synthese, 17,* 304–323.

Gauker, C. (2006). Against stepping back: A critique of contextualist approaches to the semantical paradoxes. *Journal of Philosophical Logic*, *35*, 393–422.

Glanzberg, M. (2003). A contextual-hierarchical approach to truth and the Liar paradox. *Journal of Philosophical Logic*, *33*, 27–88.

Parsons, C. (1974). The Liar paradox. *Journal of Philosophical Logic*, *3*, 381–412.

Pleitz, M. (2010). 'This sentence' is indexical. The indexical variant of the Liar paradox and McTaggart's paradox. In M. Peliš (Ed.), *The Logica yearbook 2009* (pp. 195–208). London: King's College Publications.

Prior, A. (1961). On a family of paradoxes. *Notre Dame Journal of Formal Logic*, *2*(1), 16–32.

Simmons, K. (1993). *Universality and the Liar. An essay on truth and the diagonal argument*. Cambridge: Cambridge University Press.

Simmons, K. (2007). Revenge and context. In J. Beall (Ed.), *Revenge of the Liar. New essays on the paradox* (pp. 345–367). Oxford: Oxford University Press.

Tarski, A. (1956). *Logic, semantics, metamathematics. Papers from 1923 to 1938*. Oxford: Clarendon.

Martin Pleitz
Department of Philosophy, University of Münster
Domplatz 23, D-48143 Münster, Germany
e-mail: `martinpleitz@web.de`

Sher and Shapiro on Logical Terms

Gil Sagi*

Abstract

The issue of logical terms is central to the works of Gila Sher and Stewart Shapiro on logical consequence. They each propose a formal criterion for logical terms within a model-theoretic framework, based on the idea of invariance under isomorphism. The equivalence of the two criteria, which has been overlooked until now, is proved in the paper. This provides a common ground for evaluating and comparing Sher and Shapiro's philosophical justification of their criteria.

1 Introduction

Criteria for logical terms have been a subject of interest in logic in recent years. Characterizing logical terms is especially important when assuming a condition of formality on logical consequence along Tarskian lines. For Tarski, the logical validity of an argument is determined by the forms of the sentences involved. Forms, in turn, are determined by the logical vocabulary (Tarski, 1983). So logical terms determine logical validity, and the study of logical consequence requires some account of what logical terms are.

Gila Sher and Stewart Shapiro both work in the Tarskian model-theoretic tradition, and they both address the issue of formality through the lens of logical terms. Sher and Shapiro have both formulated a formal criterion for logical terms. However, they nest their formulations in separate settings, each approaching logical consequence from a different angle.

---

*I would like to thank the participants of the Logica conference for a helpful discussion. I would also like to thank JC Beall, Ole Hjortland, Hannes Leitgeb, David Ripley, Oron Shagrir and Stewart Shapiro for comments on this paper.

Logical terms play a prominent role in Sher's Formal-Structural View of logic; her criterion, which will henceforth be referred to as *Sher's criterion*, is taken to demonstrate the notion of formality, which, for Sher, is the defining feature of logical consequence. Shapiro, on the other hand, takes logical consequence to be a cluster notion, involving modal, semantic and epistemic intuitions. The *isomorphism property*, Shapiro's formal condition on logical terms, serves as a precondition for any system meeting modal and semantic intuitions. Both writers rely strongly on isomorphism of set-theoretic structures for characterizing logical terms. However, they articulate their criteria rather differently. Sher provides a tool for assessing whether a given *term* should be admissible as a logical term, whereas Shapiro suggests a formal property of *systems* that have reasonable logical terms. A point that has been missed out in the debate so far, and can benefit the understanding of both views, is the fact that Sher's and Shapiro's criteria are essentially the same. In what follows, we prove the equivalence of Sher's criterion and Shapiro's isomorphism property, and suggest to compare their views from this common ground.

2   Definitions

2.1   Logical terms, one by one

Sher (1991, 1996) proposes a mathematical definition of logical terms, which is presented below. In Sher's approach (also known as the Tarski-Sher Approach) logical terms are terms that are *invariant under isomorphic structures*. For Sher, "Logical terms are formal in the sense of denoting properties and relations that are, roughly, intuitively structural or mathematical" (Sher, 1996, p. 668). Motivation for Sher's view can be found in Tarski's characterization of the formality of logical consequence:

> [The relation of logical consequence between a class of sentences K and a sentence X] cannot be influenced in any way by empirical knowledge, and in particular by knowledge of the objects to which the sentence X or the sentences of the class K refer (Tarski, 1983, pp. 414–415).

Thus, in Sher's view, logical terms do not distinguish between objects in the world, and this feature is explicated as invariance under isomorphic structures—logical terms are sensitive only to structure.

Sher's criterion. Formally, Sher's system extends standard classical first order logic. In her system, each logical term is a truth-functional connective, or an n-place predicate or functor of level 1 or 2. Let a model M be a pair $\langle A, I \rangle$ where A, the domain, is a nonempty set, and I is an interpretation function. Each logical term C has a corresponding function f_C such that for each model M, $f_C(M)$ is C's extension in M. Specifically:

A. If C is a first-level n-place predicate, then $f_C(M)$ is a subset of A^n.

B. If C is a first-level n-place functor, then $f_C(M)$ is a function from A^n into A.

C. If C is a second-level n-place predicate, then $f_C(M)$ is a subset of $B_1 \times \cdots \times B_n$, where for $1 \leq i \leq n$,

$$B_i = \begin{cases} A & \text{if } i(C) \text{ is an individual} \\ \wp(A^m) & \text{if } i(C) \text{ is an } m\text{-place predicate} \end{cases}$$

($i(C)$ being the ith argument of C, $\wp(A^m)$ the power set of A^m).

D. If C is a second-level n-place functor, then $f_C(M)$ is a function from $B_1 \times \cdots \times B_n$ into B_{n+1}, where for $1 \leq i \leq n+1$ B_i is defined as in (C).

In this framework, Sher contends that the following conditions are individually necessary and collectively sufficient for a term C to be admissible as a logical term:

1. (a) If C is a first-level n-place predicate, M and M' are models with universes A and A' respectively, $\langle b_1, \ldots, b_n \rangle \in A^n$, $\langle b'_1, \ldots, b'_n \rangle \in A'^m$, and the structures $\langle A, \langle b_1, \ldots, b_n \rangle \rangle$ and $\langle A', \langle b'_1, \ldots, b'_n \rangle \rangle$ are isomorphic, then $\langle b_1, \ldots, b_n \rangle \in f_C(M)$ iff $\langle b'_1, \ldots, b'_n \rangle \in f_C(M')$.

   (b) If C is a second-level n-place predicate, M and M' are models with universes A and A' respectively, $\langle D_1, \ldots, D_n \rangle \in B_1 \times \cdots \times B_n$, $\langle D'_1, \ldots, D'_n \rangle \in B'_1 \times \cdots \times B'_n$ (where for

$1 \leq i \leq n$, B_i and B_i' are as in (C) above), and the struc-
tures $\langle A, \langle D_1, \ldots, D_n \rangle \rangle$, $\langle A', \langle D_1', \ldots, D_n' \rangle \rangle$ are isomorphic,
then $\langle D_1, \ldots, D_n \rangle \in f_C(M)$ iff $\langle D_1', \ldots, D_n' \rangle \in f_C(M')$.

2. Analogously for functors (Sher, 1991, pp. 54–55).

The conditions above serve jointly as a criterion for introducing
new logical terms into a language as well as for justifying existing ones.
Standard classical logical terms (including identity) satisfy Sher's criterion of invariance under isomorphism, and so do many others, a
notable example being Mostowskian generalized quantifiers (such as
there are countably many).

2.2   Logical terms, a systemic view

An alternative approach, advocated by Shapiro, provides a more general condition on logical terms. Shapiro proposes the *isomorphism
property* as a necessary (though, perhaps, not sufficient) condition on
languages (as opposed to terms), *vis-à-vis* their logical terms, and can
be formulated as follows:

The isomorphism property (Shapiro). A language $\mathcal{L}$ (in the formalism) is said to have the isomorphism property iff for every formula
φ and models M and M' for $\mathcal{L}$ with corresponding assignments s and
s' such that $\langle M, s \rangle$ and $\langle M', s' \rangle$ are isomorphic with respect to the
nonlogical terms in φ, $\langle M, s \rangle \models \varphi$ if and only if $\langle M', s' \rangle \models \varphi$ (Shapiro,
1998, p. 151).[1]

In contrast to Sher's criterion, Shapiro takes language as a whole,
and poses the isomorphism property as a necessary condition for "any
model theory worthy of the name" (Shapiro, 1998, p. 152). As Shapiro
explains, "The isomorphism property is a manifestation of the intuition that logical truth and logical consequence should be a matter of
'form', to the extent that isomorphism preserves 'formal' features of
various models (whatever those are)" (Shapiro, 1998, pp. 151–152).

---

[1] In his (Shapiro, 1998), Shapiro does not mention assignments, but it is natural
to add them in, since a formula is evaluated by a model only under an assignment.
Alternatively, we may restrict formulas to sentences, as Shapiro does in (Shapiro,
2005). This difference doesn't have significant implications in what follows.

3   Meeting point

Indeed, Shapiro does not specify a particular logical framework as Sher does, but curiously enough, in the framework specified by Sher (classical first-order logic with first and second order logical terms) their conditions coincide. That is:

Equivalence result. Let $\mathcal{L}$ be a first order language with logical terms of level 1 or 2 (predicates and functional expressions, as in Sher's conditions A–D above), and countably many nonlogical predicates from each arity. We assume also that the model theory is a standard classical one. In particular, let each model M for $\mathcal{L}$ be a pair, $M = \langle A, I \rangle$, as specified in Sher's criterion. Then the isomorphism property holds for $\mathcal{L}$ if and only if the logical terms in $\mathcal{L}$ satisfy Sher's criterion.[2]

As shown by the equivalence result, the choice between Sher's criterion and Shapiro's isomorphism property becomes a practical matter depending on the specific task at hand. The isomorphism property has the advantage of conciseness and generality. On the other hand, if we have a language $\mathcal{L}$ for which we already know that the logical terms are well behaved in Sher and Shapiro's terms, Sher's criterion makes it easier to introduce an additional logical term C to $\mathcal{L}$. Sher's criterion allows us to check C's admissibility specifically, without having to reconsider the whole system after its addition, which the isomorphism property would require us to do. The equivalence result shows the conservativeness of systems satisfying Sher and Shapiro's criteria: adding an admissible logical term to such a system cannot result in a system that dissatisfies the criteria.[3] The equivalence result has noteworthy philosophical consequences as well, which will be suggested in the following section.

Proof of the equivalence result. For the left to right direction we need to show that if the isomorphism property holds for $\mathcal{L}$, then all its

---

[2]Note that our use of the term *logical term* is neutral and does not presuppose any criterion for admissibility.

[3]It should be noted that since Shapiro takes the isomorphism property to be a necessary, but perhaps not sufficient, condition on formal systems, he leaves the possibility of additional constraints that would rule out some terms satisfying Sher's criterion.

logical terms satisfy Sher's criterion. This is proved by constructing, for each logical term C, an appropriate formula φ and using the
isomorphism property for φ, as follows.

1. (a) Let C be an n-place first-level predicate for some $n \geq 1$,
        and assume that C is a logical term in $\mathcal{L}$. We need to show
        that it satisfies condition 1(a) in Sher's criterion.

        Let M and M' be models with domains A and A', and
        $\langle b_1, \ldots, b_n \rangle \in A^n$, $\langle b'_1, \ldots, b'_n \rangle \in A'^n$ such that the struc-
        tures $\langle A, \langle b_1, \ldots, b_n \rangle \rangle$ and $\langle A', \langle b'_1, \ldots, b'_n \rangle \rangle$ are isomorphic.
        We now define assignments s and s' for M and M' respec-
        tively, thus:

        $$s(x_i) = \begin{cases} b_i & \text{if } 1 \leq i \leq n \\ b_n & \text{if } i > n \end{cases} \qquad s'(x_i) = \begin{cases} b'_i & \text{if } 1 \leq i \leq n \\ b'_n & \text{if } i > n \end{cases}$$

        Note that the structures $\langle A, \langle b_1, \ldots, b_n \rangle, s \rangle$ and
        $\langle A', \langle b'_1, \ldots, b'_n \rangle, s' \rangle$ are isomorphic.

        Now consider the formula $\varphi = C(x_1, \ldots, x_n)$. From the
        previous claim we get that $\langle M, s \rangle$ and $\langle M', s' \rangle$ are isomor-
        phic with respect to the nonlogical terms in φ (the non-
        logical terms in φ include only variables[4]). So by the as-
        sumption that the isomorphism property holds, $\langle M, s \rangle \models$
        φ iff $\langle M', s' \rangle \models \varphi$. Therefore, by the definitions of s
        and s', $M \models C[b_1, \ldots, b_n]$ iff $M' \models C[b'_1, \ldots, b'_n]$, thus
        $\langle b_1, \ldots, b_n \rangle \in f_C(M)$ iff $\langle b'_1, \ldots, b'_n \rangle \in f_C(M')$, as required.

   (b) The case of second-level predicates is similar to that of
        first-level predicates. However, the proof in this case can-
        not proceed completely analogously to that in 1(a), since
        we do not have second-level variables in $\mathcal{L}$. We must there-
        fore employ first-level nonlogical predicates from $\mathcal{L}$ in order
        to construct the desired formula. The appropriate adjust-
        ments of the proof of 1(a) to this case are left to the reader.

---

[4]More precisely: the nonlogical terms in φ include *at most* variables. Variables can be considered as either nonlogical terms, or as neither logical nor
nonlogical—which is how Sher treats them, see (Sher, 1991, p. 84). That decision
does not affect our proof.

2. (a) Let C be an n-place first-level functional expression for some n, and assume C is a logical term in $\mathcal{L}$. Let M and M' be models with domains A and A' respectively, and assume $\langle b_1, \ldots, b_n, b_{n+1} \rangle \in A^{n+1}$, $\langle b'_1, \ldots, b'_n, b'_{n+1} \rangle \in A'^{n+1}$ are such that the structures $\langle A, \langle b_1, \ldots, b_n, b_{n+1} \rangle \rangle$ and $\langle A', \langle b'_1, \ldots, b'_n, b'_{n+1} \rangle \rangle$ are isomorphic. We need to show that $f_C(M)(b_1, \ldots, b_n) = b_{n+1}$ iff $f_C(M')(b'_1, \ldots, b'_n) = b'_{n+1}$.

As in 1(a), we define assignments s and s' for M and M' respectively:

$$s(x_i) = \begin{cases} b_i & \text{if } 1 \leq i \leq n+1 \\ b_{n+1} & \text{if } i > n+1 \end{cases}$$

$$s'(x_i) = \begin{cases} b'_i & \text{if } 1 \leq i \leq n+1 \\ b'_{n+1} & \text{if } i > n+1 \end{cases}$$

Note that the structures $\langle A, \langle b_1, \ldots, b_n, b_{n+1} \rangle, s \rangle$ and $\langle A', \langle b'_1, \ldots, b'_n, b'_{n+1} \rangle, s' \rangle$ are isomorphic.

Let R be a nonlogical binary predicate in $\mathcal{L}$. Consider now the models M_* and M'_* in which R is interpreted as identity and are otherwise identical to M and M' respectively.

Consider the formula $\varphi = R(C(x_1, \ldots, x_n), x_{n+1})$. $\langle M_*, s \rangle$ and $\langle M'_*, s' \rangle$ are isomorphic with respect to the nonlogical terms in φ (which include, at most, R and variables). Hence, by the assumption that the isomorphism property holds, $\langle M_*, s \rangle \models \varphi$ iff $\langle M'_*, s' \rangle \models \varphi$.

So by the definitions of s and s', $M_* \models \varphi[b_1, \ldots, b_n, b_{n+1}]$ iff $M'_* \models \varphi[b'_1, \ldots, b'_n, b'_{n+1}]$, i.e. $f_C(M_*)(b_1, \ldots, b_n) = b_{n+1}$ iff $f_C(M'_*)(b'_1, \ldots, b'_n) = b'_{n+1}$. Since M and M_* don't differ with respect to C, and nor do M' and M'_*, we have $f_C(M)(b_1, \ldots, b_n) = b_{n+1}$ iff $f_C(M')(b'_1, \ldots, b'_n) = b'_{n+1}$, as required.

(b) The case of second-level functional expressions falls into two cases, according to the level of their values (see condition (D) in Sher's criterion). In each case, $f_C(M)$ is a function from $B_1 \times \cdots \times B_n$ into B_{n+1}, where for $1 \leq i \leq n+1$

B_i is defined, as in (C) in Sher's criterion, to be either A or $\wp(A^m)$ for some positive m. The two cases are distinguished by what B_{n+1} is. If $B_{n+1} = A$, then the values of the function $f_C(M)$ are objects in the domain A (so in terms of its values $f_C(M)$ is like a first-level function), and a proof similar to the one given for first-level function symbols can apply (with appropriate modifications as indicated in the case of second-level predicates). If $B_{n+1} = \wp(A^m)$ for some positive m, then the values of the function $f_C(M)$ for each M are m-place relations over A. We then consider the formula that is received by applying the function symbol to a syntactically appropriate argument (an n-tuple of variables and predicates according to what $B_1, \ldots, B_n$ are), and all that to an m-tuple of variables. The details of this section of the proof are left to the reader.

3. The case of truth functional connectives is immediate.

For the right to left direction of the equivalence claim we need to show that if all the logical terms in $\mathcal{L}$ satisfy Sher's criterion, then the isomorphism property holds for $\mathcal{L}$. This is proved by induction on the complexity of φ.

A. First-level individual terms:

Lemma 1   *Let $\mathcal{L}$ be a first order language with logical terms satisfying Sher's criterion. Let t be a first order individual term in $\mathcal{L}$, let $\langle M, s \rangle$ and $\langle M', s' \rangle$ be isomorphic with respect to the nonlogical terms in t, and let F be an isomorphism from A (the domain of M) onto A' (the domain of M'), then $F(t^{\langle M,s \rangle}) = t^{\langle M',s' \rangle}$.*

Proof. By induction on the construction of t.   ■

B. We proceed to prove that Sher's criterion entails the isomorphism property by induction on the complexity of φ. The case of second-level functional expressions is left out for the sake of simplicity.

(i) Let φ be an atomic formula, $\varphi = P(t_1, \ldots, t_n)$ for some n-place predicate P and individual terms $t_1, \ldots, t_n$. Assume that $\langle M, s \rangle$ and $\langle M', s' \rangle$ are isomorphic with respect to the nonlogical terms in φ.

If P is a nonlogical term we are done, since by isomorphism and Lemma 1, $\langle M, s \rangle \models P(t_1, \ldots, t_n)$ iff $\langle M', s' \rangle \models P(t_1, \ldots, t_n)$, as required.

Now assume P is a logical term. By isomorphism and Lemma 1 the structures $\langle A, \langle t_1^{\langle M, s \rangle}, \ldots, t_n^{\langle M, s \rangle} \rangle \rangle$ and $\langle A', \langle t_1^{\langle M', s' \rangle}, \ldots, t_n^{\langle M', s' \rangle} \rangle \rangle$ are isomorphic. Therefore, by Sher's criterion, $\langle t_1^{\langle M, s \rangle}, \ldots, t_n^{\langle M, s \rangle} \rangle \in f_P(M)$ iff $\langle t_1^{\langle M', s' \rangle}, \ldots, t_n^{\langle M', s' \rangle} \rangle \in f_P(M')$, so $\langle M, s \rangle \models P(t_1, \ldots, t_n)$ iff $\langle M', s' \rangle \models P(t_1, \ldots, t_n)$, as required.

(ii) The case of truth-functional connectives is trivial.

(iii) Now assume $\varphi = Cx(\xi_1(x, x_1, \ldots, x_{k_1}), \ldots, \xi_n(x, x_1, \ldots, x_{k_n}))$ where C is an n-place second-level predicate, and for each $1 \leq i \leq n$, $\xi_i(x, x_1, \ldots, x_{k_i})$ is an individual term or a formula that contains at most $k_i + 1$ variables. Assume that $\langle M, s \rangle$ and $\langle M', s' \rangle$ are isomorphic with respect to the nonlogical terms in φ, and that A and A' are the domains of M and M' respectively.

C is a logical term, since we assumed that the second-level terms in $\mathcal{L}$ are all logical terms. (Were C allowed to be a nonlogical term, the proof would have been immediate, as in (i)).

Let us define the following objects and sets:

For each $1 \leq i \leq n$ such that ξ_i is an *individual term*, we define:

$$D_i = \xi_i(x, x_1, \ldots, x_{k_i})^{\langle M, s \rangle}$$
$$D'_i = \xi_i(x, x_1, \ldots, x_{k_i})^{\langle M', s' \rangle}$$

and for each $1 \leq i \leq n$ such that ξ_i is a *formula*, we define:

$$D_i = \{ b \in A; \; \langle M, s \rangle \models \xi_i[b, x_1, \ldots, x_{k_i}] \}$$
$$D'_i = \{ b \in A'; \; \langle M', s' \rangle \models \xi_i[b, x_1, \ldots, x_{k_i}] \}$$

Lemma 2   *For each $1 \leq i \leq n$ such that ξ_i is an individual term,*

$$F(D_i) = F(\xi_i(x, x_1, \ldots, x_{k_i})^{\langle M, s \rangle}) =$$
$$\xi_i(x, x_1, \ldots, x_{k_i})^{\langle M', s' \rangle} = D'_i,$$

where F is the given isomorphism.

Lemma 3   *For each $1 \leq i \leq n$ such that ξ_i is a formula, the given isomorphism function F induces an isomorphism from D_i onto D'_i (i.e. F restricted to D_i is an isomorphism from D_i onto D'_i).*

The proofs of the lemmas are left to the reader.

By Lemma 2 and Lemma 3, the structures $\langle A, \langle D_1, \ldots, D_n \rangle \rangle$, $\langle A', \langle D'_1, \ldots, D'_n \rangle \rangle$ are isomorphic by the same given isomorphism function F. Therefore, by Sher's criterion, $\langle D_1, \ldots, D_n \rangle \in f_C(M)$ iff $\langle D'_1, \ldots, D'_n \rangle \in f_C(M')$, i.e., by the above definition,

$$\langle M, s \rangle \models Cx(\xi_1(x, x_1, \ldots, x_{k_1}), \ldots, \xi_n(x, x_1, \ldots, x_{k_n})) \text{ iff}$$
$$\langle M', s' \rangle \models Cx(\xi_1(x, x_1, \ldots, x_{k_1}), \ldots, \xi_n(x, x_1, \ldots, x_{k_n}))$$

as required.

Comment for section (iii): For the sake of simplicity, we considered only the case where the n-place second-level predicate C applies just to one variable (as e.g. in the case of the classical quantifiers), i.e., in Sher's formulation, the case where for each $1 \leq i \leq n$, $i(C)$ is an individual or a unary predicate (cf. condition (C) in Sher's criterion). The proof easily generalizes for cases where $i(C)$ is an m-place predicate for $m \geq 1$. $\square$

4   A basis for comparison

It may seem that Sher and Shapiro work on separate projects in logic, or at least have different attitudes towards the subject. Recognizing what they share and where they diverge can thus be illuminating.

They both adhere to Tarski's pronouncement that logical consequence is *necessary* and *formal*. For both, *necessity* can be formulated as truth-preservation over all possible worlds. They both adhere to the notion of formality we stated in the introduction above: logical consequence is determined by the form of sentences. Giving essentially the same criterion for logical terms further strenghens their Tarskian brotherhood.

Given that Sher and Shapiro agree on the formal definitions, it will be interesting to evaluate their divergent philosophical doctrines in comparison to one another. Due to lack of space, I will only point out in the direction of such a comparison.

Both Sher and Shapiro invest considerable effort in explaining the role of models in their systems. In a Tarskian setting, arguments are evaluated with respect to models. An argument is taken to be logically valid if and only if it preserves truth in all models. The choice of logical terms affects the range of models that are considered—a consequence of our equivalence result is that Sher and Shapiro deal with the same range of models. Models, however, are mere mathematical objects. How can model-theory assure us that we have the right notion of logical validity, and specifically, that the arguments validated by the models are necessary and formal?

Sher and Shapiro take models to represent items in reality such that truth-preservation over these items guarentees necessity and formality. Their choice of items, however, is a point where their philosophical views differ. Shapiro takes models to represent *possible worlds under interpretations of the nonlogical vocabulary* (Shapiro, 1998), straightforwardly addressing the conditions of necessity and formality as we have formulated them. Sher, on the other hand, while agreeing on the formulations of these conditions, is reluctant to let the logical system have any reference to possible worlds. She thus takes models to represent *formally possible structures* (Sher, 1996).

An analysis and comparison of these two approaches to models and their relation to the criteria for logical terms is a possible starting point for further research.

5   Conclusion

The aim of this paper is to bring into light the connection between
Sher and Shapiro's work. Sher and Shapiro propose their criteria for
logical terms on different grounds. Shapiro looks at systems of logic
and proposes a pre-condition for them to be considered as formal,
where isomorphism is taken to preserve formal features of *models*.
Sher motivates her criterion through an analysis of what it is for a
term to be formal. By proving that Sher's criterion and Shapiro's
isomorphism property are equivalent, we provide an initial platform
for comparing their views. We further suggest to look at the roles
Sher and Shapiro assign to models in their theories as an exemplary
point of departure from the common ground.

References

Shapiro, S. (1998). Logical consequence: Models and modality. In
M. Schirn (Ed.), *The philosophy of mathematics today* (pp. 131–
156). Oxford: Oxford Univerity Press.

Shapiro, S. (2005). Logical consequence, proof theory, and model
theory. In S. Shapiro (Ed.), *The oxford handbook of philosophy of
mathematics and logic* (pp. 651–670). Oxford: Oxford Univerity
Press.

Sher, G. (1991). *The bounds of logic: a generalized viewpoint*. Cambridge, MA: MIT Press.

Sher, G. (1996). Did Tarski commit 'Tarski's fallacy'? *The Journal
of Symbolic Logic, 61*(2), 653–686.

Sher, G. (2008). Tarski's thesis. In D. Patterson (Ed.), *New essays on
tarski and philosophy* (pp. 300–339). Oxford: Oxford University
Press.

Tarski, A. (1983). On the concept of logical consequence. In J. Corcoran (Ed.), *Logic, semantics, metamathematics* (pp. 409–420).
Indianapolis: Hackett.

Tarski, A. (1986). What are logical notions? *History and Philosophy
of Logic, 7*, 143–154.

Gil Sagi
Philosophy, The Hebrew University of Jerusalem
Jerusalem, Israel
e-mail: `gilisagi@gmail.com`

Logics of Moore's Paradox

Igor Sedlár*   Juraj Podroužek

Abstract

The paper proves that the usualy mentioned conditions for Moore's paradox (i.e. factivity and consistency with positive introspection) are not necessary for MP. We demonstrate this by defining the minimal logics of MP and by proving that the logic of both versions of MP (omissive and comissive) **KM** is a proper subset of both **T** and **D4**. Moreover, we show that the logic of the comissive MP is a subset of the logic of the omissive MP **KO**. Therefore, **KO** is the minimal logic of both versions of MP. In the second half of the paper we consider a justification counterpart of **KO**, the logic **JKO**. We provide a Hilbert-style calculus for **JKO** and we define Mkrtychev models for **JKO**. We also discuss the problems of a semantics based on the more elaborate Fitting models.

1   Moore's paradox and modal logic

Moore's paradox (MP) is a famous philosophical puzzle arising from the fact that sentences of the form 'p, but X does not believe that p,' and of the form 'Not p, but X does believe that p' *might be true*, but it is *absurd* for X to assert them. Sentences of the former form are known as *omissive Moorean sentences*, sentences of the latter form are known as *comissive Moorean sentences*. Resolving MP amounts to explaining the nature of the absurdity around Moorean sentences.

---

*We would like to express our gratitude for valuable feedback to the audience at *Logica 2010*. Earlier versions of the paper have been read at Tilburg University and at the Institute of Philosophy of the Slovak Academy of Sciences. Audiences at the respective workshops have helped to improve the text considerably. The ideas underlying Section 4 were discussed with S. Artemov, to whom we are grateful for insights and encouragement.

Ever since its publication, the paradox has enjoyed notable attention, see the representative volume (Green & Williams, 2007).

One of the clearest logical renderings of the paradox comes from Hintikka's seminal book (Hintikka, 1962). If A is a sentece, we refer to the sentence 'X believes that A' (for a given X) as the *X-belief closure of A*. Hintikka demonstrates, within the framework of his logic of belief, that omissive Moorean sentences are satisfiable, but if they refer to a given X, their X-belief closures are unsatisfiable. In other words, $p \wedge \neg B_a p$ may very well be true, unlike $B_a(p \wedge \neg B_a p)$.

Hintikka's demonstration rests upon the assumption that every possible state of affairs has a doxastic alternative, and that $B_a p$ implies $B_a B_a p$.[1] We may sum it up by saying that, essentially, Hintikka demonstrated that $\neg B_a(p \wedge \neg B_a)$ is a theorem of the doxastic version of the normal modal logic **D4**.

Next, F. B. Fitch has shown that a demonstration of the unsatisfiability of $B_a(p \wedge \neg B_a)$ need not rest on Hintikka's assumptions. His famous paper (Fitch, 1963) contains a theorem to the effect that for every operator $\square$, such that $\square p \rightarrow p$ is valid, the formula $\neg\square(p \wedge \neg\square p)$ is valid. This means that $\neg\square(p \wedge \neg\square p)$ is a theorem of **T**.

These two early results are fairly general. They reveal that a version of omissive Moore's paradox holds for every operator O such that $p \wedge \neg Op$ is satisfiable, and that behaves like the box operator of some modal logic containing **D4** or **T**.[2]

It is clear that, from an abstract point of view, the situation around Moore's paradox (hereafter MP) boils down to the fact that

1. $p \wedge \neg\square p$ is satisfiable, but $\square(p \wedge \neg\square p)$ is not, or

2. $\neg p \wedge \square p$ is satisfiable, but $\square(\neg p \wedge \square p)$ is not.

We shall say that a modal logic 'is in the scope of the omissive MP' iff it satisfies the first condition, and that a logic 'is in the scope of the comissive MP' iff it satifsies the second condition.

---

[1] The latter assumption is often criticized as a plainly false claim about the nature of human belief. This criticism sometimes serves as a pretence for the complete refusal of Hintikka's solution of MP. However, it is arguable that the criticism rests upon a misunderstanding of Hintikka's treatment of the belief operator.

[2] For a discussion of similar results concerning the comissive version of the paradox, and for a survey of the scope of MP within the realm of propositional modal logics, see (Podroužek & Sedlár, 2010).

In a sense, solving MP amounts to explaining *why* the conditions hold for a given operator within a given logic. We do not attempt to do it here. Instead, we try to map the territory in which the paradox resides. What are the weakest assumptions about the □-operator that send us straight into the land of paradox? We prove later on that these are not the usually mentioned ones, *i.e.* Hintikka's or Fitch's.

Our question can be rephrased. What are the minimal logics satisfying the above conditions? In Section 2 we answer by introducing the *minimal logics of Moore's paradox* **KO**, **KC**, and **KM**. They are minimal in the sense that the only requirement concerning the □-operator is that it is subject to the omissive, the comissive, or both versions of MP, respectively.

In Section 3 we demonstrate that the logic **KM** is weaker than **T** and **D4**. The result may also be obtained by proving that, with respect to the class of all Kripke-frames, the characteristic axioms of **T** and **D4** are not consequences of $\neg\Box(p \wedge \neg\Box p)$ and $\neg\Box(\neg p \wedge \Box p)$. However, consider the logics of Moore's paradox to be a possibly interesting and usefull by-product of the modal approach to Moore's paradox.

In Section 4 the by-products begin to live a life on their own. We consider the justification counterpart of the logic **KO**, give its semantics and hint at some interesting problems.

2   The logics KO, KC, and KM

In this section we define the normal modal logics **KO**, **KC**, and **KM**. We will do this syntactically, *i.e.* by giving sets of formulas that generate these logics. Then we move on to semantical considerations and we prove the respective soundness and completeness theorems.

Note that the negation of the necessitation of the omissive moorean formula $(\neg\Box(p \wedge \neg\Box p))$ is equivalent to $\Diamond(p \to \Box p)$ and hence to

$$\Box p \to \Diamond\Box p \tag{1}$$

Note also that the negation of the necessitation of the comissive moorean formula $(\neg\Box(\neg p \wedge \Box p))$ is equivalent to $\Diamond(\Box p \to p)$ and hence to

$$\Box\Box p \to \Diamond p \tag{2}$$

We use these simpler formulas to generate the normal modal logics[3] **KO**, **KC**, and **KM**.

Definition 1

1. **KO** is generated by the formula $\Box p \rightarrow \Diamond \Box p$,

2. **KC** is generated by $\Box\Box p \rightarrow \Diamond p$,

3. **KM** is generated by the set $\{\Box p \rightarrow \Diamond \Box p, \Box\Box p \rightarrow \Diamond p\}$.

Is there a semantic way to approach these logics? Indeed, there is. Consider the following classes of Kripke frames.

Definition 2

a) $O = \{\mathfrak{F} \mid \mathfrak{F} \models \forall x \exists y(Rxy \wedge \forall z(Ryz \rightarrow Rxz))\}$

b) $C = \{\mathfrak{F} \mid \mathfrak{F} \models \forall x \exists y(Rxy \wedge \exists z(Rxz \wedge Rzy))\}$

Note that both O and C are subclasses of the class of serial frames. We will refer to the frames $\mathfrak{F} \in O$ as 'semi-transitive' frames and to frames $\mathfrak{F} \in C$ as 'semi-euclidean' frames. For a frame that is semi-transitive but not transitive, see Figure 1. For a frame that is semi-euclidean but not euclidean, see Figure 2.

The classes O and C are the key to the semantic rendering of the minimal logics of MP. In fact, the respective completeness theorems are simple corollaries of a combination of several fundamental results of modal correspondence theory. (Subsequently we build upon van Benthem's seminal book van Benthem, 1983 and upon the detailed exposition of correspondence theory in Blackburn et al., 2001) We begin with the following lemma.

Lemma 1 $\Box p \rightarrow \Diamond\Box p$ *and* $\Box\Box p \rightarrow \Diamond p$ *are Sahlqvist formulas. Moreover,*

a) $\Box p \rightarrow \Diamond\Box p$ *defines quasi-transitivity*

---

[3]Just to make sure, here are the definitions. A *normal modal logic* is any set of modal formulas that contains $\Box(p \rightarrow q) \rightarrow (\Box p \rightarrow \Box q)$ and $\Diamond p \leftrightarrow \neg\Box\neg p$, that is closed under modus ponens, uniform substitution and modal generalization (necessitation). A normal modal logic **L** is *generated* by a set of formulas Γ iff $\Gamma \subseteq L$. If **L** is generated by a singleton set $\{\phi\}$, then we say that **L** is generated by the formula ϕ. For more details see (Blackburn et al., 2001).

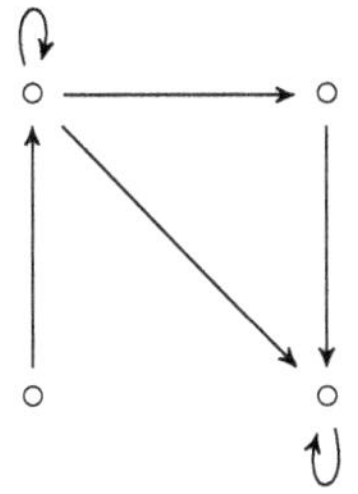

Figure 1: A quasi-transitive but non-transitive frame

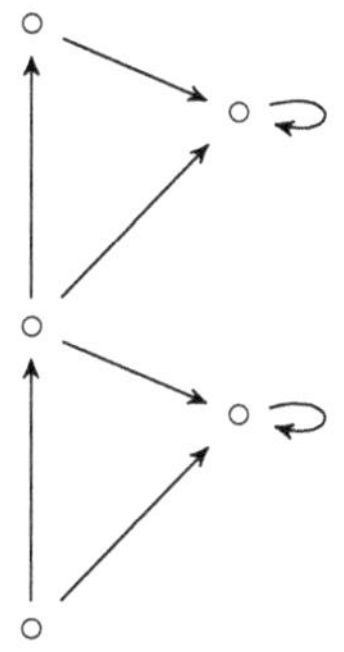

Figure 2: A quasi-euclidean but non-euclidean frame

b) $\Box\Box p \to \Diamond p$ *defines quasi-euclideanness*

Proof. For the first part, we may easily check that antecedents of both $\Box p \to \Diamond\Box p$ and $\Box\Box p \to \Diamond p$ are boxed atoms and that their consequents are positive formulas. Hence, the implications are Sahlqvist formulas. (For the respective definitions, see Blackburn et al., 2001, pp. 153–165)

For the second part, see (van Benthem, 1983) or use the Sahlqvist-van Benthem algorithm described in (Blackburn et al., 2001). It follows that the conjunction of the two formulas defines the conjunction of the two properties. $\Box$

Next, consider the following extremely useful theorem. We state it explicitly because of its central position in our completeness proof.

Theorem 2 (Sahlqvist Completeness Theorem) *Every Sahlqvist formula is canonical for the first-order property it defines. Hence, given a set of Sahlqvist axioms Σ, the logic $\mathbf{K\Sigma}$ is strongly complete with respect to the first-order class of frames defined by Σ.*

Proof. For a discussion of the theorem and its algebraic proof, see (Blackburn et al., 2001, p. 210, p. 322–325). $\Box$

Now we have everything we need for the completeness theorem.

Theorem 3 $\mathbf{KO}$ *is strongly complete with respect to the class of frames* O, $\mathbf{KC}$ *is strongly complete wrt* C, *and* $\mathbf{KM}$ *is strongly complete wrt* $\mathsf{O} \cap \mathsf{C}$.

Proof. Immediate consequence of Lemma 1 and Theorem 2. $\Box$

3   Inclusions

In this section we prove that $\mathbf{KM}$ is properly contained in both $\mathbf{T}$ and $\mathbf{D4}$. This demonstrates that the conditions given by Hintikka and Fitch are not minimal for MP. But first, we prove that $\mathbf{KC}$ is contained in $\mathbf{KO}$ and therefore $\mathbf{KO} = \mathbf{KM}$. Hence, the comissive MP is in a sense reducible to the omissive case.

Note that $\mathbf{D} \subseteq \mathbf{KO}$, since every frame in O is serial.

Lemma 4 $\mathbf{KC} \subseteq \mathbf{KO}$.

Proof. We demonstrate that $\Box\Box p \to \Diamond p \in$ **KO**.

1. $\Box p \to \Diamond \Box p$	(O axiom)
2. $\Box p \to \Diamond p$	(since **D** $\subseteq$ **KO**)
3. $\Box p \to \Diamond\Diamond p$	(1., 2.)
4. $\Box\neg p \to \Diamond\Diamond\neg p$	(3., substitution $\neg p/p$)
5. $\Box\Box p \to \Diamond p$	(4., Dual and propositional reasoning)

$\Box$

Lemma 4 states that the comissive MP is reducible to the omissive case. Hence, every operator that is subject to the omissive MP is also subject to the comissive MP. However, the converse does not hold. Thus, **KC** is a proper subset of **KO**. An obvious consequence is that **KO** = **KM**. In other words, **KO** is the minimal logic of both versions of MP.

Theorem 5 *KM $\subset$ T, D4*.

Proof. The fact that **KM** $\subseteq$ **T**, **D4** follows from earlier results by Hintikka and Fitch, and from Lemma 4. Thus we need only to demonstrate that the inclusion is proper. Consider the frame

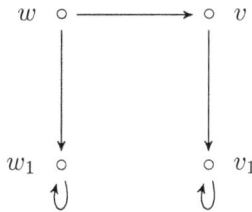

It is easy to see that the frame is semi-transitive. Build a model $\mathfrak{M}$ based on this frame by taking a valuation such that $w_1, v \Vdash p$ and $w, v_1 \nVdash p$. Hence $w \Vdash \Box p$ but $w \nVdash \Box\Box p$. Thus, $w \nVdash \Box p \to p$ and $w \nVdash \Box p \to \Box\Box p$. It follows that the T axiom and the 4 axiom are not in **KO**. The rest follows by Lemma 4. $\Box$

Theorem 5 demonstrates that the requirements proposed by Hintikka and Fitch are not the minimal requirements for MP to arise. In other words, these conditions are not necessary for MP. This means that there are non-factive operators without positive introspection that are nevertheless subject to MP. These operators need to be consistent (in the sense of $\Box p \to \Diamond p$), but consistency is not a sufficient condition for MP, as $\Box p \to \Diamond \Box p$ is not a theorem of **D**.

It seems that we have proved only that some given conditions are not the minimal necessary ones. But what are the minimal ones? The answer is trivial. A formula states a minimal necessary condition for MP iff it is equivalent (wrt the class of all frames) to the O axiom $\Box p \to \Diamond \Box p$. An interesting example of such a formula is

$$\Box \neg \Box p \to \neg \Box p \tag{3}$$

i.e. the principle of *factivity of negative introspection*. In terms of belief this means that in order to encounter MP we only need to suppose that if an agent believes that he does not believe that p, then he indeed does not believe that p.

4   A justification counterpart

This section investigates into the connections of the logics of MP with *justification logics*. Justification logics are propositional systems that, besides the usual boolean base, work with justification terms s, t, ... and operations on such terms. Hence, they allow to consider justification formulas of the form $[t]\phi$, read as 't is a justification of ϕ', where t may be considerably complex. The seminal paper on justification logic is (Artemov, 2001), for recent developments and applications to epistemology, see (Artemov, 2008) and (Artemov, 2010).

It is well known that the standard normal modal logics, such as **K**, **T**, **D**, **S4** and **S5**, have *justification counterparts*. This means, for example, that there is a justification logic **J** such that **K** proves a formula ϕ iff there is a way to replace all occurences of $\Box$ within ϕ with justification terms such that the result is provable in **J**. Results like this are known as *realization theorems* and can be found e.g. in (Artemov, 1995; Brezhnev & Kuznets, 2006; Brezhnev, 2001; Rubtsova, 2006). The central feature of the semantics of justification logics is a special rendering of justification terms. See mainly the early paper (Mkrtychev, 1997) and its elaboration (Fitting, 2005).

In this section we consider a justification counterpart of the logic **KO**, the logic **JKO**. We present a Hilbert-style calculus and a semantics for this logic based on Mkrtychev models. We conclude by pointing out some open problems and interesting questions.

First, we define the basic justification logic **J** (we follow Artemov, 2008).

Definition 3 Fix a countable set of propositional variables Φ and two disjoint sets of justification variables $Var = \{x, y, z, \ldots\}$ and constants $Cons = \{a, b, c, \ldots\}$. The set of justification terms Tm contains Var, $Const$ and every $[s + t], [s \cdot t]$ such that $s, t \in Tm$.

Axiom schemes are

1. Every propositional axiom scheme   (Prop)

2. $[s](\phi \to \psi) \to ([t]\phi \to [s \cdot t]\psi)$ (Application)

3. $([s]\phi \lor [t]\phi) \to [s + t]\phi$ (Sum)

A formula ϕ is an axiom if it is an instance of some axiom schema. Inference rules are modus ponens and

4. For every tuple of constants $(a_1, \ldots, a_n)$ and every axiom ϕ infer $[a_1] \ldots [a_n]\phi$ (R4)

We have no space here to explain the definition in detail. However, a short comment might be appropriate. Schema 2 means that the complex justification $[s \cdot t]$ justifies ψ if $[t]$ justifies ϕ and $[s]$ justifies the implication $\phi \to \psi$. Schema 3 means that $[s+t]$ justifies everything that is justified by $[s]$ or by $[t]$. The rule (R4) corresponds to the fact that any constant justifies any axiom, any constant justifies that any constant justifies any axiom, etc.

We proceed by extending **J** (corresponding to the basic normal modal logic **K**) by a justification counterpart of the schema $\Box\neg\Box\phi \to \neg\Box\phi$. We will use the following justification schema:

$$[s]\neg[s + t]\phi \to \neg[t]\phi. \tag{4}$$

The schema is an outcome of a *normal* realization of the modal schema $\Box\neg\Box\phi \to \neg\Box\phi$. This roughly means that every positive occurence of $\Box$ is realized by a complex term consisting of the terms realizing the negative occurences. For details, see (Artemov, 2001).

Note that the complex justification $s+t$ justifies every formula that is justified by s or by t (and possibly other formulas as well). Thus if $[s + t]$ does not justify ϕ then most certainly t (nor s) does not justify ϕ. But what if s justifies *the claim that* $[s + t]$ does not justify ϕ?

Intuitively, Schema 4 claims that if the fact that $s+t$ does not justify a formula ϕ is justified by s, then t does not justify ϕ either. But is this not provable already in **J**? The following derivation demonstrates that this would require the factivity schema $[t]\phi \rightarrow \phi$.

$$[s]\neg[s+t]\phi \nvdash_{\mathbf{J}} \neg[t]\phi$$

1. $[t]\phi \rightarrow [s+t]\phi$	Sum
2. $([t]\phi \rightarrow [s+t]\phi) \rightarrow (\neg[s+t]\phi \rightarrow \neg[t]\phi)$	Prop
3. $[a]([t]\phi \rightarrow [s+t]\phi)$	1., R4
4. $[b](([t]\phi \rightarrow [s+t]\phi) \rightarrow (\neg[s+t]\phi \rightarrow \neg[t]\phi))$	2., R4
5. $[b \cdot a](\neg[s+t]\phi \rightarrow \neg[t]\phi)$	3., 4., Application
6. $[s]\neg[s+t]\phi$	Assumption
7. $[(b \cdot a) \cdot s](\neg[t]\phi)$	5., 6., Application

It is not possible to infer $\neg[t]\phi$, since **J** lacks the factivity axiom, and also every restricted version of it. We may therefore understand the Schema 4 as a (very) restricted version of the factivity axiom. Now we define the Hilbert-style calculus for the logic **JKO**.

Definition 4   Axioms of **JKO** are instances of

1. Every propostitional axiom schme   (Prop)

2. $[s](\phi \rightarrow \psi) \rightarrow ([t]\phi \rightarrow [s \cdot t]\psi)$ (Application)

3. $([s]\phi \vee [t]\phi) \rightarrow [s+t]\phi$ (Sum)

4. $[s]\neg[s+t]\phi \rightarrow \neg[t]\phi$ (Restricted factivity)

Rules of inference are modus ponens and the rule (R4). A formula ϕ is **JKO**-provable ($\vdash_{\mathbf{JKO}} \phi$) iff there is a proof that contains only axioms of **JKO**, formulas derived by the inference rules and ϕ.

The proper step now would be to prove the realization theorem to the effect that **KO** proves ϕ iff **JKO** proves $(\phi)^r$ for some realization r. However, we only conjecture that the theorem holds and postpone the task of proving it to a later time.

4.1 Mkrtychev models for JKO

We now introduce the semantics of **JKO** and discuss some notable problems connected with it. Our semantics follows Mkrtychev's paper (Mkrtychev, 1997) and Kuznets' application of the technique to different justification logics in (Kuznets, 2006). Of course, our first choice was to devise a semantics based on Fitting models. However, we have encountered severe problems while trying to prove completeness via the canonical Fitting model construction. We discuss this in the next subsection.

Now to the Mkrtychev-style semantics for **JKO**.

Definition 5 A function $*$ from the set of justification terms to the power set of the set of formulas is a *proof-theorem assignment* if it satisfies the conditions

1. $\phi \to \psi \in *(s)$ and $\phi \in *(t)$ imply $\psi \in *(s \cdot t)$ ($*$ application)

2. $*(s) \cup *(t) \subseteq *(s+t)$ ($*$ sum)

3. $[a_1]\dots[a_n]\phi \in *(a)$ for every axiom ϕ, every constant a and every tuple of constants $(a_1,\dots,a_n)$ ($*$ R4)

A proof-theorem assignment is *quasi-transitive* if it in addition satisfies

4. $\neg[s+t]\phi \in *(s)$ implies $\phi \notin *(t)$ ($*$ quasi-transitivity)

A **JKO**-*model* $\mathcal{M}$ is a triple $(v, *, \models)$ where v is a propositional valuation (hence a function from Φ to $\{0,1\}$), $*$ is a quasi-transitive proof-theorem assignment, and $\models$ is a satisfaction relation defined by

1. $\mathcal{M} \models p$ iff $v(p) = 1$, for $p \in \Phi$

2. $\mathcal{M} \models \neg\phi$ iff $\mathcal{M} \nvDash \phi$

3. $\mathcal{M} \models \phi \vee \psi$ iff $\mathcal{M} \models \phi$ or $\mathcal{M} \models \psi$

4. $\mathcal{M} \models [t]\phi$ iff $\phi \in *(t)$

($\bot, \wedge, \to$ and $\leftrightarrow$ are defined in the usual way and their satisfaction-conditions are easily derived.)

A formula is *satisfied in* $\mathcal{M}$ iff $\mathcal{M} \models \phi$, ϕ is **JKO**-*valid* iff it is satisfied in every **JKO**-model. Similarly for sets of formulas.

It is plain that conditions imposed on $*$ correspond to the axioms and rules of **JKO**. Thus soundness is easily proved. The completeness proof uses a standard canonical-style argument based on the Lindenbaum Lemma.

Definition 6   Let **L** be a logic. A set of formulas Γ is called **L**-*consistent* iff there is no subset Γ' of Γ such that $\Gamma' = \{\phi_1, \ldots, \phi_n\}$ and $\vdash_{\mathbf{L}} (\phi_1 \wedge \ldots \wedge \phi_n) \to \bot$. Γ is called *maximal* **L**-*consistent* iff it is **L**-consistent and for every ψ either $\psi \in \Gamma$ or $\neg\psi \in \Gamma$.

Lemma 6 (Lindenbaum)   *If Γ is a **L**-consistent set, then there is a maximal **L**-consistent set Δ such that $\Gamma \subseteq \Delta$. Moreover, any maximal **L**-consistent set contains the axioms of **L** and is closed under the inference rules of **L**.*

The key to the completeness proof is to demonstrate that every maximal **JKO**-consistent set of formulas is satisfied in some **JKO**-model.

Lemma 7   *If Γ is a maximal **JKO**-consistent set, then there is a **JKO**-model $\mathcal{M}$ such that $\mathcal{M} \models \Gamma$.*

Proof. We construct a **JKO**-model based on the set Γ. let $*_\Gamma(t) = \{\phi \mid [t]\phi \in \Gamma\}$. It is plain that $*_\Gamma$ satisfies $*$-application, $*$-sum and $*$-R4. It also satisfies $*$-quasi-transitivity. For suppose that for some ϕ it holds that $\neg[s + t]\phi \in *(s)$. This means that $[s]\neg[s + t]\phi \in \Gamma$. Since Γ is maximal **JKO**-consistent, it contains $[s]\neg[s + t]\phi \to \neg[t]\phi$ and is closed under the inference rules of **JKO** by Lemma 6. Hence, $\neg[t]\phi \in \Gamma$. This means that $\phi \notin *_\Gamma(t)$.

Let the valuation v_Γ be defined by $v_\Gamma(p) = 1$ iff $p \in \Gamma$. Now define the model $\mathcal{M}_\Gamma$ to be the triple $(v_\Gamma, *_\Gamma, \models)$, where the coditions for $\models$ are the usual ones, see Definition 5. It is possible to demonstrate (by induction on the complexity of formulas) that for every ϕ we have $\mathcal{M}_\Gamma \models \phi$ iff $\phi \in \Gamma$. □

The completeness theorem follows as a simple consequence.

Theorem 8   *If ϕ is **JKO**-valid, then it is **JKO**-provable.*

Proof. Suppose $\nvdash_{\mathbf{JKO}} \phi$. Then the set $\{\neg\phi\}$ is **JKO**-consistent. By the Lindenbaum Lemma, extend it to a maximal **JKO**-consistent set Γ. According to Lemma 7, there is a model $\mathcal{M}_\Gamma$ such that $\mathcal{M}_\Gamma \models \Gamma$. Hence, ϕ is not **JKO**-valid. □

4.2 Problems with Fitting models

The construction of Fitting-style semantics of **JKO** has been hindered by problems concering the canonical Fitting model for **JKO**. In short, it seems that the accesibility relation R of the canonical Fitting frame is not quasi-transitive in the sense of $\forall x \in W \exists y \in W(Rxy \wedge \forall z \in W(Ryz \rightarrow Rxz))$.

Definition 7 A *Fitting-model* for **JKO** is a tuple $\mathcal{M}_F = (W, R, E, V)$ where W is a nonempty set, R is a quasi-transitive binary relation on W (i.e. $\forall x \in W \exists y \in W(Rxy \wedge \forall z \in W(Ryz \rightarrow Rxz)))$, V is a valuation and E is an evidence function from $W \times Tm$ to 2^{Fm} satisfying the conditions

1. $\phi \rightarrow \psi \in E(w, s)$ and $\phi \in E(w, t)$ imply $\psi \in E(w, s \cdot t)$

2. $E(w, s) \cup E(w, t) \subseteq E(w, s + t)$

3. $\neg[s + t]\phi \in E(w, s)$ implies $\phi \notin E(w, t)$

4. If ϕ is an axiom and a, a_1, ..., a_n are constants, then $[a_1] \ldots [a_n]\phi \in E(w, a)$ for every a and every $w \in W$.

It is clear that the conditions 1.–4. correspond to the axioms and rules of **JKO**. Hence, soundness is easily proved. But the completeness proof using the canonical model construction is problematic.[4]

Definition 8 Let Γ, Δ, ... be maximal **JKO**-consistent sets of formulas. Then

$$\Gamma^\sharp = \{\phi \mid [t]\phi \in \Gamma \text{ for some } t\}$$
$$\Gamma' = \{[t]\phi \mid [t]\phi \in \Gamma\} \text{ for some } t$$

Let the canonical W be the set of all maximal **JKO**-consistent sets of formulas. The relation R for the canonical model is defined by

$$\Gamma R \Delta \text{ iff } \Gamma^\sharp \subseteq \Delta.$$

The canonical evidence function is defined by

$$E(\Gamma, t) = \{\phi \mid [t]\phi \in \Gamma\}.$$

---

[4]The canonical model construction for justification logics is explained in detail in (Fitting, 2005).

The canonical valuation is defined by

$$\Gamma \in V(p) \text{ iff } p \in \Gamma.$$

It possible to demonstrate that the canonical frame (W, R) of the above canonical model is not quasi-transitive. Note that every quasi-transitive relation is serial. Hence to prove that the canonical R is quasi-transitive, we have to demonstrate first that for every maximal **JKO**-consistent Γ there is a maximal **JKO**-consistent Δ such that $\Gamma^\sharp \subseteq \Delta$. This amounts to demonstrating that every $\Gamma^\sharp$ is consistent (and therefore can be extended to a maximal **JKO**-consistent Δ). That this is not the case can be shown by employing the following lemma.

Lemma 9 $\nvdash_{JKO} \neg[t]\bot$.

Proof. By Theorem 8 it is sufficient to construct a Mkrtychev **JKO**-model $\mathcal{M}$ such that $\mathcal{M} \models [t]\bot$ for some t. But note that there is no contradiction in having a $*$ function with $*(x) = \{\bot\}$ even if we have $[a_1]\dots[a_n]\neg\bot \in *(a)$ for every constant a, a_1, ..., a_n and for $n \geq 0$.   □

Theorem 10 *The canonical R is not serial. Hence, it is not quasi-transitive.*

Proof. Consider the set $\Gamma_0 = \{ [x]\bot \}$. Γ_0 is consistent by Lemma 9. Extend it to a maximal **JKO**-consistent set Γ. By the definition of Γ_0 the set $\Gamma^\sharp$ contains $\bot$. Thus there is no maximal **JKO**-consistent set Δ such that $\Gamma^\sharp \subseteq \Delta$. Hence there is no Δ such that $\Gamma R \Delta$.   □

Lemma 9 implies that **JD** is not a sublogic[5] of **JKO**. This is interesting for, as we have seen in Section 3, **D** is a sublogic of **KO**. If the conjecture that **JKO** is a realization of **KO** holds, then this is somewhat surprising. This would mean that relations of inclusion between modal logics are not always preserved under realizations. Of course, this raises a general problem: Delimit the class of modal logics C such that if $\mathbf{L} \in \mathsf{C}$ and $\mathbf{L'} \subseteq \mathbf{L}$, then the $\mathbf{JL'} \subseteq \mathbf{JL}$, where **JL** and

---

[5]We say that $\mathbf{L'}$ is a sublogic of $\mathbf{L}$ iff every theorem of $\mathbf{L'}$ is provable in $\mathbf{L}$. In symbols $\mathbf{L'} \subseteq \mathbf{L}$.

$\mathbf{JL'}$ are justification counterparts of $\mathbf{L}$ and $\mathbf{L'}$ respectively (of course, here we consider only logics that have justification counterprats[6]).

Theorem 10 is similarly surprising. Note that $\mathbf{KO}$ is a *canonical logic* ('is canonical' in short) for the canonical frame for $\mathbf{KO}$ belongs to the class of frames with respect to which $\mathbf{KO}$ is sound and complete (i.e. is quasi-transitive.) Theorem 10 demonstrates that the canonical frame of $\mathbf{JKO}$ *is not* quasi-transitive. Hence it does not belong to the class of frames for the implicit modal counterpart of $\mathbf{JKO}$. This is not very usual, for the canonical frames of $\mathbf{JT}$, $\mathbf{JD}$, $\mathbf{J4}$ and the other well known justification logics *are* reflexive, serial, transitive etc. respectively.

5   Conclusion

We have seen that the usual conditions for MP (i.e. factivity and consistency with positive introspection) are not minimal for MP. We have demonstrated this by defining the minimal logics of MP and by proving that the logic of both versions of MP (omissive and comissive) $\mathbf{KM}$ is a proper subset of both $\mathbf{T}$ and $\mathbf{D4}$. Moreover, we have seen that the logic of the comissive MP is a subset of the logic of the omissive MP $\mathbf{KO}$. Therefore, $\mathbf{KO}$ is the minimal logic of both versions of MP.

In the second half of the paper we have considered a justification counterpart of $\mathbf{KO}$, the logic $\mathbf{JKO}$. We have provided a Hilbert-style calculus for $\mathbf{JKO}$ and conjectured that it is a realization of $\mathbf{KO}$. Then we defined Mkrtychev models for $\mathbf{JKO}$ and proved soundness and completeness. We also discussed the problems we have encountered while trying to devise a semantics based on the more elaborate Fitting models. It seems that these problems could be a starting point of fruitful inquiries. First, as Lemma 9 reveals, inclusions between normal modal logics are not generally preserved under realizations. For which logics is this the case? Second, as the closely related Theorem 10 shows, the canonical Fitting frames of justification counterparts of some modal logics do not belong to the class of frames for those modal logics. Again, for which logics is this the case? For now we only express hope that further research will settle these issues.

---

[6] Another general problem arises here: Delimit the class of modal logics that have explicit counterparts.

References

Artemov, S. (1995). *Operational modal logic* (Tech. Rep.). Cornell University.

Artemov, S. (2001). Explicit provability and constructive semantics. *Bulletin of Symbolic Logic*, *7*(1), 1–36.

Artemov, S. (2008). The logic of justification. *The Review of Symbolic Logic*, *1*(4), 477–513.

Artemov, S. (2010). Tracking evidence. In A. Blass, N. Dershowitz, & W. Reisig (Eds.), *Fields of logic and computation, essays dedicated to Yuri Gurevich on the occasion of his 70th birthday* (Vol. 6300, pp. 61–74). Springer.

van Benthem, J. (1983). *Modal logic and classical logic*. Neapol: Bibliopolis.

Blackburn, P., Rijke, M. de, & Venema, Y. (2001). *Modal logic*. Cambridge: Cambridge UP.

Brezhnev, V. (2001). On the logic of proofs. In *Proceedings of the sixth ESSLLI student session, 13th European Summer School in Logic, Language and Information (ESSLLI'01)*.

Brezhnev, V., & Kuznets, R. (2006). Making knowledge explicit: How hard it is. *Theoretical Computer Science*, *357*(1–3), 23–34.

Fitch, F. B. (1963). A logical analysis of some value concepts. *Journal of Symbolic Logic*, *28*, 135–142.

Fitting, M. (2005). The logic of proofs, semantically. *Annals of Pure and Applied Logic*, *132*(1), 1–25.

Green, M., & Williams, J. N. (Eds.). (2007). *Moore's paradox. New essays on belief, rationality, and the first person*. Oxford: Oxford University Press.

Hintikka, J. (1962). *Knowledge and belief*. Ithaca: Cornell UP.

Kuznets, R. (2006). Complexity of evidence-based knowledge. In *Proceedings of the workshop on Rationality and Knowledge, 18th European Summer School in Logic, Language and Information (ESSLLI'06)*.

Mkrtychev, A. (1997). Models for the logic of proofs. In S. Adian & A. Nerode (Eds.), *Logical Foundations of Computer Science, 4th international symposium, LFCS'97, Yaroslavl, Russia, July 6–12, 1997, proceedings* (Vol. 1234, pp. 266–275). Springer.

Podroužek, J., & Sedlár, I. (2010). *The scope of Moore's paradox*. (Unpublished manuscript based on a talk given at PhD's in Logic II, Tilburg, NL, 19. 2. 2010)

Rubtsova, N. (2006). On realization of S5-modality by evidence terms. *Journal of Logic and Computation*, *16*(5), 671–684.

Igor Sedlár
Department of Logic and Philosophy of Science, Comenius University
Šafárikovo námestie 6,818 01 Bratislava, Slovakia
e-mail: sedlar@fphil.uniba.sk
URL: https://sites.google.com/site/igorsedlar/

Juraj Podroužek
Institute of Philosophy, Slovak Academy of Science
Klemensova 19, 813 64 Bratislava, Slovakia
e-mail: juraj.podoruzek@gmail.com
URL: http://www.klemens.sav.sk/fiusav/

A Note on (In)Compatibility Relations

Sebastian Sequoiah-Grayson*

Abstract

Non-symmetric incompatibility relations, or non-symmetric compatibility relations, are a standard method for introducing a split negation pair on a frame. Another standard method is to reject commutation for the frame. The first task is to examine the relationship between non-symmetric incompatibility relations and non-symmetric compatibility relations and commutation failure on frames. The second task is to look at the sort of points that may constitute such frames. Taking the points as information states, what type of information states may operate in non-symmetric incompatibility/compatibility environments? The proposal made here is that databases consisting of sub-propositional information states, or data-points, are a respectable place to start.

keywords: compatibility frames, perp frames, noncommutation, split negation, van Benthem, Dunn, Restall, Wansing.

1   Introduction

Non-symmetry on an incompatibility relation $\perp$, or on a compatibility relation C, is a standard method for introducing a split negation pair

*My sincerest thanks to Greg Restall and Johan van Benthem for their good advice on many points. Many thanks also to the organisers and participants of Logica 2010. In particular, I would like to thank Marie Duzí, and Andreas Pietz, both both their comments and for their encouragement. An earlier version of this talk was presented at the ILLC in Amsterdam, and I would like to thank all of the participants, especially Dora Achourioti, Paul Dekker, and Catarina Dutilh, both for their insights and for their constructive criticism. I would also liek to take this opportunity to thank Michal Peliš and Vítek Punčochář for their tireless efforts editing the Yearbook itself.

onto a frame. Another standard method is to reject commutation for the frame. One task here is to examine the relationship between non-symmetric incompatibility relations and non-symmetric compatibility relation and commutation failure on frames. The other task is to look at the sort of points that may constitute such frames.

In section 2 we introduce compatibility frames and a split negation pair defined in terms of them. In section 3 we introduce perp frames and the resulting defined split negation pair. In section 4 we introduce noncommuting frames and the correspondingly defined split negation pair. In section 5 we translate the ternary relation R of substructural frame semantics into its operational semantics counterpart. In section 6 we look at the connections between a split negations pair defined in terms of a noncommuting frame, and in terms of C and $\perp$. In section 7 we lay out the relationships between (non)commutation on R and (a)symmetry on C and $\perp$. In section 8 we examine cases of non-symmetric (in)compatibility. In section 9 we precisify the conceptual framework underpinning much of what has transpired in sections 2–8, and consider its limits with respect to various database types.

2 Compatibility frames

A compatibility frame (Restall, 2000) $\mathbf{F_C}$ is a triple $\langle S, \sqsubseteq, C \rangle$. S is a set of information states x, y, z, ... $\sqsubseteq$ is a partial order of informational development/inclusion such that $x \sqsubseteq y$ is taken to mean that the information carried by y is a development of the information carried by x. C is a binary compatibility relation where $x \, C \, y$ means that the information carried by x is compatible with the information carried by y. If $x \, C \, y$ then there is no information not carried by x that is carried by y. C interacts with $\sqsubseteq$ in the following manner:

$$\text{If } x \sqsubseteq y \text{ and } y \, C \, z, \text{ then } x \, C \, z. \tag{1}$$

$$\text{If } x \sqsubseteq z \text{ and } y \, C \, z, \text{ then } y \, C \, x. \tag{2}$$

In what follows, we will often write $x, y, \ldots \in \mathbf{F_X}$ as shorthand for $x, y, \ldots \in S$ where $S \in \mathbf{F_X}$. Reading $x \Vdash A$ as "the information state x carries information of type A", in this case we can give the following evaluation conditions for our a split negation pair in terms of C as follows:

$$x \Vdash \sim A \text{ } \textit{iff} \text{ for each } y \in \mathbf{F_C} \text{ s.t. } x \text{ } C \text{ } y, \text{ } y \nVdash A. \tag{3}$$

$$x \Vdash \neg A \text{ } \textit{iff} \text{ for each } y \in \mathbf{F_C} \text{ s.t. } y \text{ } C \text{ } x, \text{ } y \nVdash A. \tag{4}$$

Importantly, our split negation pair is preserved only under the assumption that C is non-symmetric. That is, that:

$$x \text{ } C \text{ } y \nRightarrow y \text{ } C \text{ } x \tag{5}$$

Question 1: How plausible an assumption is (5)?

There is no correct answer to this until one has specified the type of information carried by the states $x, y, z, \ldots \in S$, which is just to say until one has specified the type of information denoted by A. We have said nothing about this yet. We will have much to say about this later, but before then we need to lay out the details at work in *incompatibility frames*.

3   Perp frames

A perp frame (or incompatibility frame, Dunn, 1994, 1996) $\mathbf{F}_\perp$ is a triple $\langle S, \sqsubseteq, \perp \rangle$ where S and $\sqsubseteq$ are as in the compatibility frame above. $\perp$ is a binary incompatibility relation such that $x \perp y$ means that the information carried by x is incompatible with the information carried by y. In this case, we may define a split negation pair as follows:

$$\sim A := \{X : A \perp X\} \tag{6}$$

$$\neg A := \{X : X \perp A\} \tag{7}$$

$\perp$ interacts with $\sqsubseteq$ in the following manner:

$$\text{If } x \sqsubseteq y \text{ and } z \perp x, \text{ then } z \perp y. \tag{8}$$

$$\text{If } x \sqsubseteq y \text{ and } x \perp z, \text{ then } y \perp z. \tag{9}$$

In this case we can give the following evaluation conditions for our a split negation pair in terms of $\perp$ as follows:

$$x \Vdash \sim A \text{ } \textit{iff} \text{ for each } y \in \mathbf{F}_\perp \text{ s.t. } y \Vdash A, \text{ } x \perp y. \tag{10}$$

$$x \Vdash \neg A \text{ } \textit{iff} \text{ for each } y \in \mathbf{F}_\perp \text{ s.t. } y \Vdash A, \text{ } y \perp x. \tag{11}$$

As with C, our split negation is preserved only under the assumption that $\perp$ is non-symmetric. That is, that:

$$x \perp y \nRightarrow y \perp x \tag{12}$$

Question 2: How plausible an assumption is (12)?

Question 3: What is the relationship between Question 1 and Question 2?

The answer to Question 2 turns on the very same issues as does the answer to the same question put to C, namely Question 1. This is in itself the answer to Question 3. Questions 1 and 2 turn on the same issues because C and $\perp$ are complements:

$$x \; C \; y \; \textit{iff} \; \text{not} \; x \perp y \tag{13}$$

$$x \perp y \; \textit{iff} \; \text{not} \; x \; C \; y \tag{14}$$

Their equivalent definitions of negation are most easily appreciated in the context of a split negation pair defined in terms of *noncommuting frames*.

4   Noncommuting frames

A noncommuting frame $\mathbf{F_{nc}}$ is a triple $\langle S, \sqsubseteq, R \rangle$ where S and $\sqsubseteq$ are as above, and R is a noncomutting ternary relation on the frame such that:

$$Rxyz \nRightarrow Ryxz \tag{15}$$

Via the addition of a double implication pair $\langle \rightarrow, \leftarrow \rangle$ and bottom constant $\mathbf{0}$, we can define our split negation pair as follows:

$$\sim A := A \rightarrow \mathbf{0} \tag{16}$$

$$\neg A := \mathbf{0} \leftarrow A \tag{17}$$

The evaluation conditions for our connectives and constant are given by the following:

$$x \Vdash A \rightarrow B \text{ iff for all } y, z \in \mathbf{F_{nc}} \text{ s.t. } Rxyz, \text{ if } y \Vdash A \text{ then } z \Vdash B.$$
(18)

$$x \Vdash B \leftarrow A \text{ iff for all } y, z \in \mathbf{F_{nc}} \text{ s.t. } Ryxz, \text{ if } y \Vdash A \text{ then } z \Vdash B.$$
(19)

$$x \Vdash \mathbf{0} \text{ for no } x \in \mathbf{F}.$$
(20)

In this case, we can give the evaluation conditions for our split negation pair via the following:

$$x \Vdash {\sim}A[A \rightarrow \mathbf{0}] \text{ iff for all } y, z \in \mathbf{F_{nc}} \text{ s.t. } Rxyz, \text{ if } y \Vdash A \text{ then } z \Vdash \mathbf{0}.$$
(21)

$$x \Vdash \neg A[\mathbf{0} \leftarrow A] \text{ iff for all } y, z \in \mathbf{F_{nc}} \text{ s.t. } Ryxz, \text{ if } y \Vdash A \text{ then } z \Vdash \mathbf{0}.$$
(22)

In $\mathbf{F_{nc}}$ our split negation pair is only preserved on account of its noncommuting behaviour. Were commutation to be present, then we would loose our split negation.

Question 4: How plausible an assumption is (15)?

Question 5: What is the relationship between Questions 4, 1, and 2?

The answer to this Question 4 turns on the same issues as does the answer to Questions 1 and 2 above. However, the answer to Question 5 is not the same as the answer to Question 4. Although there is a relationship between (in)compatibility non-symmetry and commutation failure, these are not equivalent. Getting clear on all of this is most easily done after interpreting R in robustly informational terms via an operational semantics.

5 Operational semantics and R

We can make the operation encoded by the ternary relation under an informational interpretation clearer via a rewriting in terms of an operational semantics. The point of an operational semantics is that it makes the semantic operations explicit.

We start by rewriting the ternary relation R in terms of $\sqsubseteq$ and a binary composition operator $\bullet$ that operates on members of S. $\bullet$ is the semantic counterpart to fusion, or intensional\multiplicative conjunction, $\otimes$. Our rewrite comes out as the following:

$$Rxyz := x \bullet y \sqsubseteq z \qquad (23)$$

We read our operationalised ternary relation as something like "the combination of the information carried by x with the information carried by y develops into the information carried by z". In this case, we may rewrite (15) as:

$$x \bullet y \nRightarrow y \bullet x \qquad (24)$$

In other words, commutation failure corresponds to directional combination. We may also rewrite (18) and (19) as the following:

$$x \Vdash A \to B \text{ iff for all } y, z \in \mathbf{F_{nc}} \text{ s.t. } x \bullet y \sqsubseteq z, \text{ if } y \Vdash A \text{ then } z \Vdash B. \qquad (25)$$

$$x \Vdash B \leftarrow A \text{ iff for all } y, z \in \mathbf{F_{nc}} \text{ s.t. } y \bullet x \sqsubseteq z, \text{ if } y \Vdash A \text{ then } z \Vdash B. \qquad (26)$$

Hence (21) and (22) get rewritten as:

$$x \Vdash {\sim}A[A \to \mathbf{0}]$$
$$\textit{iff} \text{ for all } y, z \in \mathbf{F_{nc}} \text{ s.t. } x \bullet y \sqsubseteq z, \text{ if } y \Vdash A \text{ then } z \Vdash \mathbf{0}. \quad (27)$$

$$x \Vdash \neg A[\mathbf{0} \leftarrow A] \textit{ iff}$$
$$\text{for all } y, z \in \mathbf{F_{nc}} \text{ s.t. } y \bullet x \sqsubseteq z, \text{ if } y \Vdash A \text{ then } z \Vdash \mathbf{0}. \quad (28)$$

We will refer to the ternary relation under its operationalised definition for much of what follows, as it brings to the foreground the very compatibility and incompatibility properties that we are interested in. Now we can clarify the relationship between non-symmetry on C and $\perp$ and noncommutation on R, namely Question 5 from section 4 above.

6 C, $\perp$, and R

Where we have $\sim A := A \to \mathbf{0}$ and $\neg A := \mathbf{0} \leftarrow A$, C and $\perp$ are closely related to R. In this case, we have it that:

$$x \, C \, y \ \textit{iff} \ \exists z(x \bullet y \sqsubseteq z) \tag{29}$$

$$y \, C \, x \ \textit{iff} \ \exists z(y \bullet x \sqsubseteq z) \tag{30}$$

$$x \perp y \ \textit{iff} \ -\exists z(x \bullet y \sqsubseteq z) \tag{31}$$

$$y \perp x \ \textit{iff} \ -\exists z(y \bullet x \sqsubseteq z) \tag{32}$$

(29)–(32) make sense. (29) and (30) state that two information states are compatible *iff* there is a further information state resulting from their combination, while (31) and (32) state that two information states are incompatible *iff* there is no information resulting from their combination. Given this, we can rewrite (3) and (4) as the following:

$$x \Vdash \sim A \ \textit{iff} \ \text{for each} \ y \in \mathbf{F_C} \ \text{s.t.} \ x \bullet y \sqsubseteq z, \ y \nVdash A. \tag{33}$$

$$x \Vdash \neg A \ \textit{iff} \ \text{for each} \ y \in \mathbf{F_C} \ \text{s.t.} \ y \bullet x \sqsubseteq z, \ y \nVdash A. \tag{34}$$

We can also rewrite (10) and (11) as:

$$x \Vdash \sim A \ \textit{iff} \ \text{for each} \ y \in \mathbf{F_\perp} \ \text{s.t.} \ y \Vdash A, \ -\exists z(x \bullet y \sqsubseteq z). \tag{35}$$

$$x \Vdash \neg A \ \textit{iff} \ \text{for each} \ y \in \mathbf{F_\perp} \ \text{s.t.} \ y \Vdash A, \ -\exists z(y \bullet x \sqsubseteq z). \tag{36}$$

(33) and (35) are equivalent. They tell us the same thing, that information states carrying information of type $\sim A$ can never be applied to an information state carrying information of type A, as such a process is "informationally redundant". Similarly, (34) and (36) are equivalent. They tell us the same thing, that information states carrying information of type $\neg A$ can never have an information state carrying information of type A applied to it, again due to procedural redundancy from the perspective of information generation. These "procedural redundancy" interpretations are the very meanings of our split negation pair that are directly encoded by our definitions (16) and (17). The evaluation conditions given by (27) and (28) bear this out. Abstracting across directional distinctions, $A \to \mathbf{0}$ and $\mathbf{0} \leftarrow A$ are simply functions which, by definition, do not return outputs when

given information of type A as an input. This is due to null-output **0** holding nowhere via (20). Before we do move on to accounting for this direction difference, we need to get a little clearer on the the constraints between commutation on R and symmetry on C and $\perp$.

7   Commutation on R and symmetry on $C(\perp)$

If R is commutative then $C(\perp)$ is symmetric:

$$(x \bullet y \sqsubseteq x \rightarrow y \bullet x \sqsubseteq z) \Rightarrow (x \ C \ y \rightarrow y \ C \ x) \qquad (37)$$

(37) hold because of the following (via (29) and (30)):

$$(x \bullet y \sqsubseteq x \rightarrow y \bullet x \sqsubseteq z) \Rightarrow \exists z(x \bullet y \sqsubseteq x) \rightarrow \exists z(y \bullet x \sqsubseteq z) \qquad (38)$$

The constraint in (38) holds only in the left-to-right direction, as there is no constraint from symmetry on $C(\perp)$ to commutation on R. We can bring this out by revealing the suppressed universal quantifiers and scope markers in (38) in order to get the following:

$$\forall x \forall y (\forall z(x \bullet y \sqsubseteq x \rightarrow y \bullet x \sqsubseteq z)) \Rightarrow (\exists z(x \bullet y \sqsubseteq x) \rightarrow \exists z(y \bullet x \sqsubseteq z)) \qquad (39)$$

This is just to say that we can have symmetry on $C(\perp)$ without full commutation on R, which is as we would expect. A neat example form Greg Restall (in conversation) bears this out: Define R on the integers by setting $Rxyz$ iff $x - y = z$. In this case, it will always be the case that $\exists z Rxyz$, hence $\exists z Rxyz \rightarrow \exists z Ryxz$. However, we have $R101$ (since $1 - 1 = 0$), but not $R011$ (since $0 - 1 \neq 1$), hence we do not have it that $Rxyz \rightarrow Ryzx$ for all x. There will always be such an x, hence we have symmetry, although the relation will not always be true of it, hence we do not have full commutation.

Since there is no in general constraint from symmetry on $C(\perp)$ to full commutation on R, we know that there is no in general constraint from commutation failure on R to symmetry failure on $C(\perp)$. However, we do know that there is an in general constraint from symmetry failure on $C(\perp)$ to commutation failure on R (simply from (37) via logic).

8 Examples of non-symmetry on $C(\bot)$

Now we are in a healthy position to examine the questions put at the end of the first three sections. Q1 and Q2 are equivalent on account of compatibility and incompatibility being equivalent. Due to (37), we know that if we can give examples of symmetry failure on $C(\bot)$, we will have examples of commutation failure for free.

The thing is this: how to make sense of the non-symmetry of $x \, C \, y$ (and hence of $x \perp y$). In (Wansing, 2001), Wansing states that "Obviously, [a split negation pair] coincide[s] if the not implausible assumption is made that incompatibility is a symmetric relation." With a very small amount of license, we will take this comment of Wansing's to be equivalent to the claim that non-symmetry on (in)compatibility *is* an implausible assumption.

The non-symmetry of the relation appears implausible because of the assumption that the information-types carried by the information states is tenseless, propositional information. Although this is standard procedure for elementary extensional logic, it is obviously not a constraint insofar as logical modeling is concerned.

The information carried by the information states may by of any type we stipulate, and some of these types will not support compatibility\incompatibility-symmetry. Still at the propositional level, consider tense-specific action-based propositions such as *Sebastian opens the fridge door* and *Sebastian retrieves a beer from the Fridge*. The compatibility of these informational tokens, and hence of the information states that carry them, is order-sensitive—I do not get my beer without first opening the fridge door.

However, in 2 we stated that $x \Vdash A$ should be read as "the information state x carries information of type A". Note we did not state "the information *that* A. That is, we have left the way open for processing on sub-propositional information types. The $C(\bot)$ asymmetries of sub-propositional information types from Categorical Grammar are suitable robust.

Directional information compatibility (or lack of it) is the mark of much of the informational behaviour of natural languages. Take an intransitive verb such as 'skips', and a noun such as 'Friederike'. 'Freiderike skips' is well-formed, whilst 'skips Friederike' is not. In this case, 'Friederike' is compatible with 'skips', however 'skips' is not compatible with 'Friederike'. The compatibility here is directional,

insofar is information generation is concerned. Trivially, it is also the
case that we get an non-symmetry of incompatibility here, as 'skips'
is incompatible with 'Friederike, but 'Friederike' is not incompatible
with 'skips'.

We can bring all of this together with operational semantics and
multiple-typing. That is, 'skips' is information of type $\neg n := \mathbf{0} \leftarrow n$,
as well as of type $n \rightarrow s$. That is, if you give 'skips' to the right
hand side of a noun, you get a sentence as the informational output,
as in (41), but if you give it to the left hand side of a noun, you do
not get any informational output, as marked by (40):

$$x \Vdash \neg n \, [\mathbf{0} \leftarrow n] \; \textit{iff} \text{ for all } y, z \text{ s.t. } y \bullet x \sqsubseteq z, \text{ if } y \Vdash n \text{ then } z \Vdash \mathbf{0}. \tag{40}$$

$$x \Vdash n \rightarrow s \; \textit{iff} \text{ for all } y, z \in \mathbf{F} \text{ s.t. } x \bullet y \sqsubseteq z, \text{ if } y \Vdash n \text{ then } z \Vdash s. \tag{41}$$

For information compatibility\incompatibility that runs in the other
direction, take as an example the adjective 'happy', and the noun
'Frederike' again. Here, 'happy Friederike' is well-formed, whilst
'Friederike happy' is not. In this case, 'Friederike' is incompatible
with 'happy', but 'happy' is not incompatible with 'Friederike'. Equivalently, 'happy' is compatible with 'Friederike', but 'Friederike' is not
compatible with 'happy'.

With operational semantics and multiple-typing again, we have
'happy' as type $\sim n := n \rightarrow \mathbf{0}$ and type $n \leftarrow n$. That is, if you give
'happy' to the left side of a noun, you get a complex noun phrase as
the informational output, as in (43), and if you give 'happy' to the
right hand side of a noun, you do not get any informational output,
as in (42):

In explicit information processing terms, the types corresponding
to 'happy' have the following frame conditions:

$$x \Vdash \, \sim n \, [n \rightarrow \mathbf{0}] \; \textit{iff} \text{ for all } y, z \text{ s.t. } x \bullet y \sqsubseteq z, \text{ if } y \Vdash n \text{ then } z \Vdash \mathbf{0}. \tag{42}$$

$$x \Vdash n \leftarrow n \; \textit{iff} \text{ for all } y, z \in \mathbf{F} \text{ s.t. } y \bullet x \sqsubseteq z, \text{ if } y \Vdash n \text{ then } z \Vdash n. \tag{43}$$

Some of the most dramatic examples of $C(\bot)$ non-symmetry then,
are the already well-known non-commuting properties of natural language.

9 Conclusion

Concrete examples of $C(\perp)$ non-symmetry are to be found in the dynamic behaviour of the sub-propositional information states underpinning natural language semantics. There is a philosophical aside that some might think threatens at this point, a purported worry about whether or not sub-propositional information really is information in any substantive sense. The concern here is that a necessary condition on information is that is be propositional. That is, are the information states in our frames above really information states, or merely data points?

There are two points to be made at this stage. The first is that metaphysical concerns such as this are in every way operationally irrelevant to the issue at hand. The second is that such concerns start to look merely terminological under a sufficient amount of light—whether you choose to call the the points data points or information states is largely a matter of taste (see Sequoiah-Grayson, 2007). The substantive issues concern the logical properties and behaviours of the models concerned, and these remain happily independent of baptisms.

Proceeding in terms of information states, the crucial moves for bringing non-symmetry on $C(\perp)$ into line with categorical grammar and natural language semantics have been to operationalise the ternary relation R in our various Kripke frames, parsing them in robust information-processing terms. By laying out the relationship between commutation failure on these frames and $C(\perp)$ non-symmetry, we then have a complete link to an information-processing parsing of (in)compatibility between information states. The example of sub-propositional information-processing can be pushed along a little further.

In our operationalised frames above, commutation-failure between particular information states is guaranteed precisely *because* these states are incompatible. We can think of the information states carrying the conditional information types as carrying functional information types, and of the information states carrying the antecedent information types as carrying input information types. In this case, compatibility between the relevant information states means that processing on these states guarantees output success. By contrast, incompatibility between the relevant information states means that processing on these states guarantees output failure. This is a good conceptual framework to use as a heuristic insofar as appreciating why it is that non-symmetry on C amounts to non-symmetry on $\bot$, and *vice versa*. Analysing logical formulas in terms of output success\failure has its proto-manifestation in (Groenendijk & Stokof, 1991), and is further developed for various propositional and sub-propositional database types in (Sequoiah-Grayson, n.d.-a, n.d.-b, 2009).

A natural language lexicon is just one database amongst many, both propositional and sub-propositional. Given the simplicity of the operational semantics, we should expect there to be a wide variety of database types to which such an analysis may be put. However, it is not obvious, indeed far from it, that *all* database types will be "cooperative". A group of agents in a communicative setting is a database if anything is. Dynamic Epistemic and Public Announcement logics (see Baltag, Moss, & Solecki, 1998; Baltag & Smetts, 2008; van Benthem, 2009; van Ditmarsh, van der Hoek, & Kooi, 2008) have been successful in capturing a great deal of the dynamic logical behaviour of such databases. In (van Benthem, 2010), van Benthem makes a strong case for there being *in principle* reasons to suspect that an operational analysis built on the back of categorical grammars breaks down when confronted with the subtleties at work in multi-agent scenarios such as these. At present this remains an open, and interesting problem.

References

Baltag, A., Moss, L., & Solecki, S. (1998). The logic of Public Announcements, Common Knowledge and Private Suspicions. In *Proceedings of Tark'98* (pp. 43–56). Morgan Kaufmann Publishers.

Baltag, A., & Smetts, S. (2008). The Logic of Conditional Doxastic Actions. In R. van Rooij & K. Apt (Eds.), *Texts in logic and games*. Amsterdam University Press.

van Benthem, J. (2009). *Logical Dynamics of Information and Interaction*. manuscript.

van Benthem, J. (2010). *Categorical versus Modal Information Theory*. forthcoming in *Linguistic Analysis*.

van Ditmarsh, H., van der Hoek, W., & Kooi, B. (2008). *Dynamic Epistemic Logic*. Springer.

Dunn, J. M. (1994). Start and Perp: Two Treatments of Negation. In J. E. Tomberlin (Ed.), *Philosophical perspectives* (Vol. 7, pp. 331–357).

Dunn, J. M. (1996). Generalised Ortho Negation. In H. Wansing (Ed.), *Negation: A notion in focus* (pp. 3–26). Berlin: Walter de Gruyter.

Groenendijk, J., & Stokof, M. (1991). Dynamic Predicate Logic. *Linguistics and Philosophy*, *14*, 33–100.

Restall, G. (2000). Defining Double Negation Elimination. *Logic Journal of the IGPL*, *8*(6), 853–860.

Sequoiah-Grayson, S. (n.d.-a). *Epistemic Closure and Commuting, Nonassociating Residuated Strucrures*. forthcoming in *Synthese*.

Sequoiah-Grayson, S. (n.d.-b). *Lambek Calculi with 0 and Test–Failure in DPL*. forthcoming in *Linguistic Analysis*.

Sequoiah-Grayson, S. (2007). The Metaphilosophy of Information. *Minds and Machines*, *17*(3), 331–344.

Sequoiah-Grayson, S. (2009). Dynamic Negation and Negative Information. *Review of Symbolic Logic*, *2*(1), 233–248.

Wansing, H. (2001). Negation. In L. Goble (Ed.), (pp. 415–436).

Sebastian Sequoiah-Grayson

Postdoctoral Research Fellow, Department of Theoretical Philosophy, University of Groningen—The Netherlands.

Senior Research Associate, IEG—Computing Laboratory, University of Oxford.

e-mail: s.sequoiah-grayson@rug.nl
URL: http://logic.tsd.net.au

Quine's other Way Out

Hartley Slater

Abstract

It is shown that, on the traditional, grammatical notion of a predicate as the remainder of a sentence once the subject term has been removed, there is no problem with Russell's Paradox, or comparable paradoxes such as Grelling's, and the Paradox of Predication. The standard formal ban on substituting predicates involving free variables into schemas where those variables would become bound is enough to prevent the standard paradoxes from developing. The re-arrangements required in the foundations of Set Theory to incorporate this insight are then discussed, and the consequences for the closely related matters Diagonalisation, and Cantor's Theorem explained.

1

I have pointed out in several places before (Slater, 2004, 2005, 2007) that the Fregean tradition mixed up predicates with the forms of sentences. A predicate (in the old, and, outside of Logic books, still current sense) is a proper part of a sentence: it is that part of a sentence that remains after the subject is removed. Thus commonly, in English, the predicate is the latter part of a sentence, the part that follows the subject that commonly comes first. In this way the predicate in 'x is not a member of x' is 'is not a member of x', and the subject is the 'x' that has then been removed. On the other hand the form of the whole sentence is '(1) is not a member of (1)', and this has been thought of as a kind of 'predicate', following Frege. On this variant understanding of 'predicate' there is also a different understanding of 'subject'. A subject in this alternative sense is not what is maybe at the start of a sentence, but becomes a term or expression that may recur throughout the sentence. Thus if '(1) is not a member of (1)' is taken as the 'predicate' in 'x is not a member of x', then 'x' becomes the 'subject' in this second sense, because it replaces '(1)' at all occurrences, not just at the start.

The distinction enables us to see that something different is said of a and of b when, for example, we say of each that he shaves himself. For what is then predicated of each does not have the verbal form '(1) shaves (1)', but simply 'shaves himself', and the 'himself' has a variable referent, dependent on its contextual antecedent. So different properties are attributed to a and to b: the property of shaving a in the one case, and the property of shaving b in the other. Of course, all those who shave themselves might still contingently share a further property, and so form a set of those who have that property, and who, incidentally, are all of those who shave themselves, as when they are all together in a room: $(x)(Rx \equiv Sxx)$. But there is no necessity that there is such an 'R' for all 'S', i.e. there is no logical equivalent of 'Sxx' of the form 'Rx' in general.

The point resolves a number of puzzles that have bedevilled twentieth century logic. For, in connection with Grelling's Paradox, a problem arises when we use such a word as 'heterological' for what 'x' is when 'x' does not apply to 'x'. For then the variable within the (old-style) predicate 'does not apply to 'x'' is obscured, since such words are properly used only for constant predicates. If instead we use 'not self applicable', the variable nature of the predicate is more apparent, although we still might forget that substituting 'not self applicable' for 'x' in:

'x' is not self applicable if and only if 'x' is not x,

means substituting it for 'self' as well as 'x', since there are four references to 'x' in the statement, and not just three. Substituting 'not self applicable' ('NSA') for 'x' in this statement does not lead to

'NSA' is NSA if and only if 'NSA' is not NSA,

but to

'NSA' is not 'NSA' applicable if and only if 'NSA' is not NSA,

which is unexceptionable.

The same applies to Russell's Paradox, the Paradox of Predication, and other forms of Grelling's Paradox. For, notoriously, if we try to represent 'x is not a member of itself' as 'x is a member of R' for some fixed 'R', then a contradiction ensues. But none does if we respect the variable nature of 'itself'. What x is necessarily a member of, for instance, if it is not a member of itself, is its complement. But 'its complement' contains the contextual element 'its', and so in

> x is a member of its complement if and only if x is not a member of x,

$(x \in x' \equiv x \notin x)$ substitution of 'its complement' ('IC') for 'x' leads not to the contradictory

> IC is a member of IC if and only if IC is not a member of IC,

$(x' \in x' \equiv x' \notin x')$ but to the unexceptionable

> IC is a member of ICs complement if and only if IC is not a member of IC,

$(x' \in x \equiv x' \notin x')$ once one remembers that there is a variable item in the predicate 'is a member of its complement'.

In the Paradox of Predication the concern is with 'x is a property it does not possess', or 'x is a property but does not possess that property', i.e. '$(\exists P)(x = P \,\&\, \neg(x \text{ has } P))$'. But this is '$x = P^* \,\&\, \neg(x \text{ has } P^*)$' with $P^* = \varepsilon P(x = P \,\&\, \neg(x \text{ has } P))$ in the epsilon reduction (see Slater, 2006, especially sections 8 and 9), which clearly shows that the property attributed to x in the (old-style) predicate is not constant, but varies with x. Likewise with Grelling's Paradox in the form ''x' does not possess the property it expresses', or ''x' expresses but does not possess a certain property' i.e. '$(\exists P)($'x' expresses $P \,\&\, \neg($'x' has $P))$'.

2

However, it has recently come to my attention that there is another way of obtaining this conclusion using a standard feature of formal logic (c.f. Slater, 2010). For if the substituted 'F' in the naïve abstraction scheme

$$(\exists y)(x)(x \text{ is a member of } y \equiv Fx),$$

had to be a predicate in the old style, then the substitution of 'is not a member of x' for 'F' would violate a formal restriction. If one tried to derive Russell's Paradox from this abstraction scheme by substituting the predicate 'is not a member of x' for 'F', to get 'x is not a member of x' for 'Fx', then this would violate the restriction that variables free in a predicate must not be such as to be captured by quantifiers in the scheme into which the predicate is substituted (c.f. Quine, 1959, p. 141). For the variable 'x' in 'is not a member of x' would become bound by the quantifier '(x)'.

There is no problem with introducing occurrences of other variables in the substituted predicate, but there is a quite general problem with bringing in a variable free in the substituted predicate that would be bound in the scheme it is substituted into. In an example from Quine, consider the substitution of 'Gx' for 'F' in

$$Fy \supset (\exists x)Fx.$$

This implication is formally valid, so the given substitution is improper since it would yield

$$Gxy \supset (\exists x)Gxx,$$

which is invalid (c.f. Quine, 1959, p. 144).

Quine himself overlooked the way this point provides a way out from Russell's Paradox. That was no doubt because the novel Fregean grammar was burnt well into him. In the way Fregeans think of it, it is quite proper that, in the scheme of naive abstraction, '$F(1)$' be replaced by '(1) is not a member of (1)', to yield

$$(\exists y)(x)(x \text{ is a member of } y \equiv x \text{ is not a member of } x).$$

Putting it this way, one is using Quine's device of 'placeholders' to indicate the argument-places of '$F(1)$'. The point to note is that the complex 'predicate' (strictly 'form of a sentence') that then replaces '$F(1)$' does not contain any occurrences of 'x', hence the above bar on capturing seemingly does not apply. Fregeans would think of themselves as substituting '(1) is not a member of (1)' not for 'Fx' but for '$F(1)$', where the argument places marked by '(1)' are filled by whatever fills the argument place of '$F(1)$'—in the above case 'x'.

But if we keep to the traditional notion of predicate as the remainder of a sentence after the removal of (in English) the first occurrence of its subject, then clearly Quine's restriction will enable us to escape the paradox that results from the Fregean way of looking at the matter. More exactly, it will enable us to escape from paradox with any substitution into the abstraction scheme

$$(\exists y)(x)(x \text{ is a member of } y \equiv Fx),$$

that does not violate the above bar on capturing. For the further point that needs to be made is that that does not preclude having further abstraction schemes applying when there is reflexivity in the predicate. There is no problem with replacing the 'F' above with any constant, old style predicate, or even such a predicate involving another variable, like 'Rz'. But being unable to replace the above 'F' with 'Rx' leaves us with the need for an abstraction scheme applicable when 'Rxx' is on the right hand side. That is no problem, however, since the way to handle relations quite generally, and so equally when the subject is repeated, is to bring in sets of ordered pairs.

If a is shaving a, then, as before, a has the property of shaving a (also the property of being shaved by a). But a also stands in a relation to himself: he and himself form a shaving (i.e. shaver-shaved) pair. Moreover, if a is shaving a, and b is shaving b, then the *same* relation is involved—a relation that a holds with himself, and b holds with *himself* (sic, notice the change of referent); and that relation is not a specifically 'reflexive relation', since it is the same relation that a would have with b, if a were shaving b. Thus quite generally,

$$(\exists y)(x)(z)(\langle z, x \rangle \text{ is a member of } y \equiv Rzx),$$

and the same y is involved if $z = x$, even though y is not just a set of ordered pairs whose members, in each case, are the same. But, in the particular case

$$(\exists y)(x)(\langle x, x \rangle \text{ is a member of } y \equiv x \text{ is not a member of } x),$$

we only get, on substitution,

$$\langle y, y \rangle \text{ is a member of } y \equiv y \text{ is not a member of } y,$$

which is not a contradiction.

Surprisingly, therefore, we must conclude that, from a traditional perspective, Frege got into his problem with Russell's Paradox through forgetting the applicability of the elementary notion of 'being free for' to the case.

3

Of course the elementary, and rather banal principles above are the basis for more interesting and complicated results, since once sets for elementary predicates are defined, those for non-elementary predicates can be constructed out of them by standard set-theoretic processes.

Thus, for a start, from the Abstraction Axiom (given Extensionality, which ensures uniqueness of referent for the epsilon term) we can invariably write '$\{\, x : Px \,\}$' or '$\varepsilon z(y)(y \in z \equiv Py)$' for the set of Ps (where the (old-style predicate) 'P' in 'Py' contains no occurrence of 'y'). Repeated variables in a relation must be handled differently, as above, but since 'x is P and x is Q' is the same as 'x is P and Q' ('x is R to y, and x is S to y' is the same as 'x is R and S to y' etc.), the repeated variables in conjunctions like '$Px \,\&\, Qx$' can be handled using the normal definition of set intersection. Thus:

$$\{\, x : Px \,\&\, Qx \,\} = \{\, x : x \in (\{\, y : Py \,\} \cap \{\, y : Qy \,\}) \,\}.$$

Likewise with the union of two sets, and the complement of a set:

$$\{\, x : Px \lor Qx \,\} = \{\, x : x \in (\{\, y : Py \,\} \cup \{\, y : Qy \,\}) \,\};$$
$$\{\, x : \neg Px \,\} = \{\, x : x \notin \{\, y : Py \,\} \,\}.$$

The null set can then be defined as the intersection of $\{\, x : Fx \,\}$ and $\{\, x : \neg Fx \,\}$ (for any 'F'), i.e.,

$$\emptyset = \varepsilon y(x)(x \in y \equiv Fx) \cap \varepsilon y(x)(x \in y \equiv \neg Fx),$$

and the universal set likewise as the union of $\{\, x : Fx \,\}$ and $\{\, x : \neg Fx \,\}$ (for any 'F').

As for the standard axioms of Set Theory, the present approach has the advantage of making most of them redundant. Thus the Axiom of Regularity is not required since there is nothing suspect about expressions like '$x \in x$', and their more complex kin. Some of the functions of the Axiom of Choice are taken over by the properties of epsilon terms (as in Bernays' formulation of Set Theory, see Bernays, 1968). For

$$(\exists x)(x \in y) \equiv \varepsilon x(x \in y) \in y,$$

and so the appropriate epsilon term always provides a selection from a non-empty set. The Power Set Axiom follows using Abstraction on the definition of a subset, for given

$$y \subset x \equiv (z)(z \in y \supset z \in x),$$

and

$$(\exists z)(t)(t \in z \equiv t \subset x),$$

then one can always define

$$\Psi x = \{\, y : y \subset x \,\} \quad (= \varepsilon z(t)(t \in z \equiv t \subset x)).$$

The Axiom of Pairs is now an immediate inference from Abstraction and the definition of the union of two sets, since

$$(\exists y)(x)(x \in y \equiv x = z),$$

yields

$$x = z \equiv x \in \{\, t : t = z \,\},$$

and so it follows, using the process of (finite) set union above, that

$$(\exists y)(x)(x \in y \equiv x = z \lor x = t).$$

The Axiom of Separation in the form

$$(\exists y)(x)(x \in y \equiv (x \in z \,\&\, Px)),$$

(where 'Px' is as before) is now an immediate inference from Abstraction and the definition of the intersection of two sets. But the Axiom of Separation is standardly expressed using, in place of 'Px', a formula in which 'x' might occur any number of times. So that will not follow in the present case, without a series of further assumptions like

$$(y)(\exists x)(t)(t \in x \equiv \langle t, t \rangle \in y).$$

This (and its kin with larger ordered sets) clearly holds if the set corresponding to 'y' is finite, and so can be listed and not just known descriptively, i.e. 'intensionally'. But it cannot hold in general, since it is just this kind of assumption, we now see, that generates Russell's Paradox.

So care must be taken with, for instance, such equivalences as

$$Rss \equiv (\exists t)(s = t \ \& \ Rtt).$$

The R.H.S. here looks like it might be of the required constant form 'Ps', and so the further assumption above may seem to be automatically satisfied. Thus 's shaves himself' is equivalent to 's is someone who shaves himself' and the predicate 'is someone who shaves himself' might seem to have a constant sense. The subject-predicate structure of the R.H.S., however, is more fully displayed in its epsilon equivalent:

$$s = t^* \ \& \ Rt^*t^*,$$

where $t^* = \varepsilon t(s = t \ \& \ Rtt)$. So in the old-style predicate in question (i.e. the portion of this last expression after the initial 's') there are again further occurrences of the subject, making the referent of the pronoun 'someone' in 'is someone who shaves himself' not constant, but a function of the subject the predicate is applied to. So while there is a constant syntactic predicate, the epsilon analysis reveals it expresses a variable property, as with 'shaves himself', 'does not apply to itself' etc.

What the above equivalence does logically ensure is that something of the following form is provable:

$$(t)(\langle t, t\rangle \in x \equiv \langle t, t, t, t\rangle \in y).$$

But with 'Rss' as 's shaves himself' then, as before, only contingently (and thus only with a finite set being involved) could there be a z such that

$$(t)(t \in z \equiv \langle t, t\rangle \in x).$$

So 'Separation' in its traditional form is not automatically guaranteed, and that also means that the Axiom of Choice, on which the full form of Separation is clearly based must not be assumed in general. For one moves in the further assumptions above from a set of ordered sets with iterated members to a set that selects just one member from each of those ordered sets (the problem being not in the selection of the various members, but in whether there is a set of all those selected when it would have to be given descriptively, or 'intensionally', being infinite). That leaves Abstraction, and Extensionality as the only two set-theoretic principles that are totally justified.

4

There are consequences for the understanding of Diagonalisation, of course, with which Russell's Paradox is closely related. For what has been called 'Cantor's Theorem' seems to show that the power set of any set has greater cardinality than the set itself. If 'x' ranges over members of a set, and 'S_x' over correlated subsets of the set, then Cantor argued that

$$(x)(x \in S_y \equiv x \notin S_x),$$

must define a further subset, S_y, i.e. the 'y' cannot name a member of the set. But if that was so then it would follow that there could be no universal set. For each of its subsets would have to be a member of it, being sets, making the cardinality of the universal set at least as great as the cardinality of its power set—a contradiction. But a universal set is easily defined, as before. So there must be something wrong with Cantor's argument, and what is wrong is now easy to diagnose. For what is true, for a start, is merely that

$$(\exists z)(x)(\langle x, x \rangle \in z \equiv x \notin S_x),$$

so Cantor needed a further premise

$$(\exists y)(t)(t \in y \equiv \langle t, t \rangle \in \varepsilon z(x)(\langle x, x \rangle \in z \equiv x \notin S_x)),$$

to establish that his 'theorem' held in general.

Likewise with other forms of 'Cantor's Theorem'. For given a defined sequence of functions of one variable, $f_x(y)$, onto $(0, 1)$ then

$$f_r(x) = 1 - f_x(x)$$

will define a different function, i.e. the 'r' will not be one of the 'x's. But in the extreme case, where the sequence contains *all* functions of one variable onto $(0, 1)$, evidently no new function can be defined in this way, without contradiction. So it is not just that there might be something like a 'non-recursive function' in such a case beyond recursive ones. What there is in the extreme case is a sequence of functions without *any* definable function of one variable generating $f_x(x)$, because from the index 'x' there is no definable function generating the function with that index, and so no 'F' such that $f_x(x) = F(x)$.

The situation, in other words (c.f. Slater, 2000, pp. 94–95), parallels that for computable functions. For while all computable functions of one numerical variable onto $(0,1)$ are enumerable, there is no way to specifically enumerate just those that have completely defined values (i.e. which are not just partial but total functions), otherwise the halting problem would be solved. Hence the ordinal numbers of those functions that are total, although denumerable, are not enumerable. There is, in other words, a further kind of expression, which is like that for a binary 'decimal' except certain places are undefined. These expressions are enumerable, but diagonalisation does not produce a further one of them, since neither $f_m(n)$, nor $1 - f_n(n)$ need equal anything. Amongst the functions which generate these expressions are all the total functions of one variable, but we cannot, in general, determine which these functions are. Even if $f_m(n)$ is total, which function it is is only determinable from its ordinal place amongst all the computable functions of one variable, not from its ordinal place amongst the total functions of this sort, with the result that, if the latter is 'm', then $f_m(n)$ is not a calculable function of m. Of course, if one *specifies* a sequence just of total functions that makes it the case that which function is the mth in that sequence is determinable from m, and $1 - f_n(n)$ will then be a further, distinct total function of n. But it is only the specification of such a sequence which makes $f_m(n)$ a function both of m and of n, and so there is no further diagonal function in an unspecified case, much as there was no diagonal set in the extreme case before.

References

Bernays, P. (1968). *Axiomatic set theory.* Amsterdam: North-Holland Pub. Co. (with an historical introduction by A.A. Fraenkel)

Quine, W. (1959). *Methods of Logic.* New York: Holt, Rinehart and Winston. (Rev. Ed.)

Slater, B. H. (1973). Is 'heterological' heterological? *Mind, 82*(327), 439–440.

Slater, B. H. (2000). The Uniform Solution. In *LOGICA Yearbook 1999.* Prague: Czech Academy of Sciences.

Slater, B. H. (2004). A Poor Concept Script. *The Australasian Journal*

of Logic, *2*, 44–55. Available from `http://www.philosophy`
`.unimelb.edu.au/ajl/2004/`
Slater, B. H. (2005). Choice and Logic. *Journal of Philosophical Logic*, *43*, 207–216.
Slater, B. H. (2006). Epsilon Calculi. *Logic Journal of the IGPL*, *14*(4), 535–590.
Slater, B. H. (2007). Logic and Grammar. *Ratio, XX*, 206–218.
Slater, B. H. (2010). Quine's other way out. *The Reasoner*, *4*(3), 37–38.

Barry Hartley Slater
Honorary Senior Research Fellow, Philosophy M207, UNIVERSITY OF WESTERN AUSTRALIA
35 Stirling Highway, Crawley, WA 6009
tel: +61 8 6488 2165
fax: +61 8 6488 1182
e-mail: `slaterbh@cyllene.uwa.edu.au`
URL: `http://www.philosophy.uwa.edu.au/about/staff/hartley_slater`

Making it More Explicit

Friedrich Slivovsky

Abstract

Robert Brandom's inferentialist philosophy involves an understanding of logic as rooted in norms governing linguistic practice. We present his account of the latter as "deontic scorekeeping" games and argue that it is at odds with his logical semantics based on the notion of incompatibility. We further contend that it does not adequately capture the distinction between inference and consequence. As an alternative, we present an interpretation of Lorenzen-style dialogue games in terms of Brandom's pragmatist categories. We close with a short discussion of the difficulties and potential merits of this approach.

1   Motivation and outline

Formal (model theoretic) semantics in the Tarski-Montague tradition gets off the ground by associating individual variables and constants with objects and sentences with truth values. The meaning of (subsentential) expressions generally is then identified with their contribution in fixing the truth values of sentences they occur in. The resulting semantic contents correspond to relations on the domain of objects. In a philosophical context, these are often imbued with ontological significance, interpreted as as what is "out there", as properties *in* the world.

On this understanding of language, issues concerning the meaning of a particular (type of) expression take the form of questions about what it *represents*, about what property or object it stands for. Grasp of meaning is cashed in as a matter of correct representation, and proper use manifests in inferences that preserve truth. This being so, truth and representation appear as the basic semantic notions congenial to the model theoretic paradigm.

By contrast, Pittsburgh philosopher Robert Brandom, in a number of seminal contributions (Brandom, 1994, 2000, 2008), developed a theory that appeals to *inference* as a semantic primitive. Inverting the above order of explanation, he sets out from the *use* of language in order to account for its semantics, arguing that the meaning of a locution should be understood as its role, its function within linguistic practice.

We will argue that, in this respect, his own *incompatibility semantics* for logical connectives (Brandom, 2008) is found wanting. Although the semantic primitive he appeals to has a straightforward pragmatic correlate, the notion of consequence built on top of it cannot be reconciled with his analysis of linguistic social practice as presented in (Brandom, 1994). It does not tell us how it would be correct for speakers to make inferences based on logical structure, if inference is understood according to Brandom's formal account of the "game of giving and asking for reasons" (henceforth *GOGAR*). We will argue that ideas from disjunctive logic programming and Lorenzen-style dialogue games (Lorenzen, 1960) could provide the basic building blocks for an alternative semantics that makes the connection to pragmatics more palpable.

The paper is structured as follows. We will begin with a coarse summary of Brandom's "inferentialism", followed by an outline of his incompatibility semantics. Against this background, we will spell out the criticism hinted at above by arguing that the notion of "incompatibility-entailment" cannot be integrated with his formal description of GOGAR. We further claim that Brandom's definition of inferential relations based on GOGAR undermines the distinction between inference and consequence. The remainder of the paper will delineate a different strategy of welding together the pragmatic and semantic dimensions, one that involves an interpretation of Lorenzen-style dialogue games as "deontic scorekeeping practices" similar to GOGAR. It is our belief that this provides a formal basis for exploring the implications of Brandom's views on logic.

2   Brandom's inferentialism

In (Brandom, 1994), Robert Brandom developed a theory of (conceptual) content that takes inference as a semantic primitive and accounts for the representational dimension of language (as well as truth-talk) in terms of it. This *inferentialism* is supplemented with an intricate story that explains how inferential relations are instituted by norms implicit in an abstract linguistic social practice he refers to as the "game of giving and asking for reasons"—a kind of normative functionalism that identifies the content of an expression with its role in language use.

More specifically, according to Brandom, this role corresponds to an assertion's potential in serving as reasons for other assertions. *Inference*, rather than *representation*, provides the cornerstone of his semantic theory. Now, what does it mean in practice for one thing to serve as reason for another? What kinds of *doings* can be thought of as the making of *inferences*? Brandom's answer appeals to the irreducibly *normative* dimension of talk about meaning. He suggests that asserting a proposition involves the undertaking of a *commitment*: a commitment to assert further propositions (that follow from it) and to vindicate one's *entitlement* to it (by asserting propositions it follows from). For instance, whoever asserts "this patch of grass is green" is implicitly committed to asserting "this patch of grass is colored." On the other hand, one can secure (defeasible) entitlement to the claim "Tweety can fly" based on the assertion of "Tweety is a bird." According to Brandom, making inferences consists in moving from one assertion to the next in a manner that preserves commitment and entitlement. Practices of reasoning are practices that institute these *deontic* statuses of commitment and entitlement which effectively encode proprieties of inference. In order for individuals to be able to competently engage in such language games, they need to keep track of what they and their fellow interlocutors are committed and entitled to. In general, the consequences of any particular claim depend on the context in which it is uttered and on what additional premises are available to draw inferences from. Brandom suggests that these facts concerning who is committed and entitled to what can be thought of as deontic *score* in the language-game—linguistic practice thus can be said to involve a kind of *deontic scorekeeping*.

2.1 Deontic scorekeeping

The interplay of inferential relations and the attribution of deontic status by speakers is illustrated in Brandom's schematic account of de-

ontic scorekeeping practices (Brandom, 1994, p. 190). It ties together
the two main components of his project, viz. his semantic inferentialism and his normative pragmatism, by characterizing the structure a
practice must exhibit in order for it to underwrite the kind of inferential relations that are constitutive of semantic content. According to
Brandom, only practices that match this description can be thought
of as properly "linguistic" practices. Strictly speaking, only the per-
formances of individuals engaged in such practices can be made sense
of as *inferring* and *asserting*.

What do scorekeeping practices look like in detail? As indicated
above, every speaker is supposed to keep track of what their fellow
interlocutors are committed and entitled to. They therefore associate
with every one of them two sets of propositions representing commitments and entitlements, respectively.

The content of a proposition p is modeled on the effect an asserting
of p has on this deontic score, which can be characterized along three
dimensions in which it might relate to some proposition q:

- It could be that whoever is committed to p is thereby committed
  to q (in Brandom's terminology, q is a *commitment-preserving
  or committive consequence* of p).

- Someone undertaking a commitment to p may thus become entitled to q (q is an *entitlement-preserving or permissive consequence* of p).

- Finally, commitment to p might preclude entitlement to q, and
  vice versa (p and q are *incompatible*).

Brandom's description of the update process triggered by B's asserting
the proposition p formally captures these relations. It involves several
steps:

1. First, p is added to the the list of commitments C_B attributed
   to B.

2. C_B is then closed with respect to commitment-preserving inferential relations.

3. Entitlements E_B associated with B might have to be revoked in
   the light of incompatibilities arising from the updated set C_B.

For instance, if $q \in E_B$ is incompatible with p, q will be removed from E_B.

4. Finally, E_B is closed under commitment-preserving inferential relations, and permissive consequences of $E_B \cap C_B$ (where not defeated by incompatibilities) are added to E_B.

It should be noted that Brandom makes no claim that this procedure is consonant with findings about how individuals *de facto* keep track of deontic score. The above is intended as a characterization of what is minimally required from scorekeepers to deploy the three kinds of semantic relations mentioned earlier. At the same time, changes at the semantics/pragmatics interface have repercussions for the inferentialist project as a whole, and different ways of filling in the details may entail different understandings of inferential relations. Further below, we will discuss a variant of the update process to deal with inferences whose conclusions are sets of disjuncts.

2.2 Incompatibility semantics

In a more recent work, Brandom develops a logical semantics based on the notion of incompatibility (Brandom, 2008). He defines a relation of logical consequence as follows: "*p incompatibility-entails q* just in case everything incompatible with q is incompatible with p" (p. 121). Formally:

Definition 1 $X \models Y$ iff $\bigcap_{p \in Y} I(p) \subseteq I(X)$,

where I maps a set of propositions to the sets of sentences it is incompatible with. (Note that $\{x_1, \ldots, x_n\} \models \{y_1, \ldots, y_m\}$ is supposed to be read as conjunctive on the left and disjunctive on the right.) Without going into too much detail, the idea is to interpret every set of sentences X in terms of $I(X)$. Having motivated this approach, Brandom goes on to introducing logical connectives in a way that in the end yields classical logic.

He stresses that he arrives at this result without recourse to *truth* or *bivalence*, notions that are usually thought of as the essence of classical logic (Brandom, 2008, p. 173). Moreover, Brandom claims that incompatibility semantics is *not compositional* and yet fully *recursive*, in the sense that logically complex claims can be computed from those associated with strictly less complex ones.[1] According to Brandom this implies the in-principle-learnability of holistic semantics (as entailed by inferentialism), against arguments to the contrary by notorious critics such as Jerry Fodor (Fodor & Lepore, 2007).

3   Scorekeeping revisited

Given Brandom's pragmatist ambitions, one would expect that incompatibility semantics is consistent with his account of deontic scorekeeping practices. But although there the notion of *incompatibility* is given a pragmatic interpretation, no mention is made of *incompatibility-entailment*. Where *should* incompatibility-entailment figure in the update process? Fortunately, Brandom provides a hint elsewhere: he suggests that permissive, committive, and incompatibility-entailment form a hierarchy, where the order is from weakest to strongest (compare how entitlements are closed wrt. committive consequence upon updating).

> It can be argued on relatively general grounds that these three sorts of inferential consequence relation can be ranked strictly by their strength: all incompatibility-entailments are commitment-preserving, and all commitment-preserving inferences are entitlement-preserving. (Brandom, 2000, p. 195)

At a minimum, incompatibility-entailment should be treated in parallel with committive entailment: scorekeepers ought to close commitments under incompatibility-entailment. But here they run into a problem. Since incompatibility-entailment is disjunctive on the right, in general p will incompatibility-entail a set of propositions q_1 *or* q_2 ... *or* q_n. Which of these disjuncts, if any, should be added to a set of commitments upon closing under (committive) consequence? It is not obvious how scorekeepers are supposed to deal with this, and as far as we are aware, Brandom himself has no solution on offer. What seems clear, however, is that you can't just throw the set of disjuncts together into a single proposition q and leave the problem to material inferences involving q to sort out. For the purposes of logic, the

---

[1] Although Brandom's interpretation of this as "reducibility" is questionable (Fermüller, 2010).

restriction to one-place conclusions may well lead to very different results. Moreover, this sort of treatment is at odds very generally with Brandom's construal of the role of logic. It is part and parcel of his theory that logical locutions serve the purpose of making explicit inferential proprieties underwritten by GOGAR. For example, $p_1 \wedge p_2 \wedge p_3 \rightarrow q$ expresses the fact that whoever is committed to all of p_1, p_2, and p_3 is thereby committed to q, viz. that any score containing the former must also contain the latter. Likewise, sentences of the form $p \rightarrow q_1 \vee q_2 \vee q_3$ should tell us something about scorekeeping practice.

Here, we suggest a treatment of disjunction that meets this condition: we take consequence relations with disjunctive right hand sides as a given and revise the notions of deontic score and scorekeeping accordingly. In general, a committive-inferential relation will take the form $p_1 \wedge \cdots \wedge p_n \rightarrow q_1 \vee \cdots \vee q_m$. We submit that *(disjunctive) logic programming* could provide formal tools to model scorekeeping involving such relations. In a nutshell, logic programs consist of *rules*, where each rule can be further decomposed into a head and a body, effectively corresponding to the conclusion and premise of an inference (rules for which the body is empty are called *facts*). One way of assigning semantics to logic programs is through so-called stable models or answer sets (Gelfond & Lifschitz, 1988). One can think of these as sets of propositions that are consistent and closed with respect to the application of rules. That is, whenever they contain all propositions in the body of a rule, they must contain at least one of the propositions in its head as well (in particular, it must contain all facts).

We can apply this to deontic scorekeeping in a straightforward manner. Instead of maintaining for each interlocutor a set of commitments that is closed under inference, one takes an entire logic program as what they are committed to, with facts roughly corresponding to those commitments that have been undertaken publicly, through overt assertions. This introduces a certain amount of indeterminacy to the concept of commitment: in general, there will be multiple (in presence of negation, even mutually incompatible) sets of propositions an interlocutor could be interpreted as committed to.

This could provide the means to characterize different scorekeeping strategies, according to what set of commitments (if any) are attributed to an interlocutor. For instance, the commitments a scorekeeper attributes to herself could serve as a point of reference. In a cooperative setting, a scorekeeper would then strive for a maximum of agreement with the commitments attributed to an interlocutor. Under adverse circumstances, she might be more interested in ascribing commitments incompatible with her own.

3.1 Inference and consequence

Scorekeeping as rendered thus far has the flavor of a single-player game. Speakers publicly make assertions, but draw their conclusions in private. The implicit attribution of score in Brandom's model has the undesirable effect that it partially undermines the distinction between *inference* and *consequence*. Recall that committive-inferential relations $p \to q$ are underwritten by scorekeeping proprieties: whenever someone is committed to p, she is also committed to q. Now, although the formal model draws a distinction between (local) inferences and the (global) consequence relation they induce, the predicate *treats q as a committive consequence of p* does not. If the disposition to attribute q whenever p is attributed is taken as a definition of what it means for scorekeepers to make inferences, the latter are already assumed to be consequence relations. It does not allow for scorekeepers to treat $p \to q$ and $q \to r$ as correct inferences without treating $p \to r$ as correct.[2]

The distinction can be properly made if deontic score is attributed in a way that is more transparent to interlocutors. Scorekeepers should make explicit the individual inference steps that lead to them to attribute a (consequential) commitment. In fact, we intend to go a step further and conceive scorekeeping as an interactive process of drawing out the implicit consequences from a set of commitments. On this idealized view, not just assertions but all operations on deontic score are public. The difference between inference and consequence then emerges as that between individual "explicating" moves and what can be made explicit in principle within this process.

---

[2]One might object that what is at stake is what they *ought to* do, in contrast to how score is *actually* kept out in the wild. But invoking the normative/factual distinction is beside the point if what is needed is a minimal descriptive account of scorekeeping practices. It should not be part of such an account that an individual's dispositions to treat moves as correct be *closed* under committive/permissive consequence.

We contend that *logical dialogue games* could provide the formal basis for this sort of understanding of deontic scorekeeping, as suggested already in (Fermüller, 2010).

4   Dialogue games in a nutshell

Logical dialogue games are due to Paul Lorenzen (Lorenzen, 1960). Originally driven by concerns with his earlier operative semantics for logics, he proposed to identify the meaning of a sentence with an adversarial dialogue about that sentence. Such dialogues consist in an exchange of statements between two players, the proponent **P** and the opponent **O**, with the intended reading that **P** puts forward an initial sentence and tries to respond to criticism from **O**.

The metaphor underlying Lorenzen's dialogues is simple but suggestive: argumentation is a game, and valid arguments are those that the **P** knows how to win. Along these lines, dialogues offer a game theoretic definition of logical consequence. $\Gamma \models \Delta$ iff **P** has a *winning strategy* for the dialogue about Δ where **O** initially grants the sentences in Γ.[3]

We now give an informal review of the general structure of dialogue games. At each state of the game, we associate (multi)sets Δ and Γ of formulae with **P** and **O**, respectively. The moves available to players classify either as *attacks* or as *defenses*. An attack involves choosing a formula associated with one's dialogue partner and, sometimes, adding a formula to one's own set of formulae. Attacks on a formula ϕ can be answered to by defenses that add a formula (usually a subformula of ϕ) to the defending player's set of formulae and usually remove ϕ in turn. However, instead of defending, players may choose to initiate an attack themselves, for instance by calling into question premises of the original attack.

The particular form assumed by these moves depends on the topmost connective of the formula that is attacked or defended. It is set down for each logical constant in the so-called *particle rules*. The table below summarizes the particle rules for Lorenzen's original game for intuitionistic logic (if **X** is the proponent, **Y** is the opponent, and vice versa):

---

[3] In the most general setting, both Γ and Δ will be finite multisets of formulae.

X:	attack by **Y**	defense by **X**
$A \wedge B$	left? or right? (**Y** chooses)	A or B, accordingly
$A \vee B$	?	A or B (**X** chooses)
$A \supset B$	A	B

Note that particle rules alone do not fully specify dialogue games. They must be paired with *structural rules* that establish winning conditions, the succession of players' moves, and so forth. When combined with a specific set of structural rules, the above particle rules provide a dialogue game characterization of intuitionistic logic (Felscher, 1986).

Notably, the same set of particle rules can give rise to different logics depending on what structural rules it is paired with. Shahid Rahman and others have developed Lorenzen's approach into a framework characterizing a great variety of logics (Rückert, 2001). As part of a parallel tradition Robin Giles described a game for Lukasziewicz logic based on dialogue games (Giles, 1974). Generalization to other fuzzy logics followed (Fermüller, 2008).

Beyond these technical results, it has been claimed that dialogue games can contribute to foundational debates in logic. Regarded an abstraction of ordinary linguistic practice, they are supposed to justify the rules associated with logical connectives and clarify logical concepts in general (Lorenz, 1968). However, critics like Wilfrid Hodges (Hodges, 2001) have pointed out that these attempts are plagued by a blatant sort of circularity: individuals must be assumed to be in possession of logical concepts before they can engage in such dialogues. He argues that dialogue games (and the motivation of players) cannot be made sense of in isolation from intentions to capture logical notions. For instance, Hodges points out that the labeling of moves as "attacks" and "defenses" is highly metaphorical and, depending on context, less antagonistic readings are more intelligible. In short, the current understanding of dialogues does not live up to these foundationalist ambitions. The interpretation as deontic scorekeeping games sketched below is intended as a first step towards a philosophically robust reading of dialogue games.

5   Dialogues as deontic scorekeeping

Our suggestion, then, is to read dialogue games as deontic scorekeeping practices. The sets of sentences associated with **O** and **P** should accordingly be understood as sets of commitments. Where Lorenzen distinguishes attacks and defenses, we understand both as operations on deontic score. The combination of an attack and the corresponding defense explicates the score according to a committive-inferential relation. In sequences of such moves, **P** and **O** make explicit the tacit consequences of the initial score in a step-by-step fashion.

Dialogues allow us to distinguish two levels of inferential relations, replacing Brandom's indiscriminate definition of committive consequence in terms of closure of deontic score:

- Individual explicating moves underwrite "basic" (committive-) inferential relations (similar to rules in a logic program). As with Lorenzen's constructive rule for disjunction, these may involve assistance from the interlocutor to whom the commitment is attributed.

- The consequences Δ of a set of initial commitments Γ which can be made explicit in the course of a full dialogue (in the sense that **P** has a winning strategy for the dialogue about Δ where **O** grants Γ at the beginning).[4]

One manifest difficulty with this sort of interpretation is Lorenzen's rule for implication, for two reasons. First, explicating Y's score involves the undertaking of a commitment (the premise) by X which is generally not a committive consequence of her own score. Second, strictly speaking Y would have to be committed to the premise herself in order for X to have a point in attributing commitment to the conclusion. How do we deal with this? As a preliminary answer, we are inclined to think of X's "attack" on an implication $A \supset B$ as involving two steps. The first is an assertion of A, adding A to her set of commitments. Y then inherits commitment to A by default: unless Y has reason to doubt X's entitlement to A, she must consider A as a premise in inferences.

---

[4]We stress that this distinction assumes a purely heuristic status. E.g. we do not subscribe to an interpretation of atomic moves as corresponding to analytic inferences.

On this reading, however, it is not obvious why the undertaking of new commitments should be restricted to the above case. And in fact, allowing **O** to assert further sentences should not change her winning strategies (ignoring the possibility of dragging out games she would have already lost otherwise). According to Lorenzen's rules, **P** wins if and only if the dialogue is finite and **P** makes the final move. If anything, **O**'s undertaking new commitments makes the game harder for her to win. Likewise, any additional sentence **P** asserts merely gives **O** additional opportunities to attack. However, the intended interpretation of dialogues as the scorekeeping practices obviates the strict distinction between attacks and defenses. In general, we would like to discard Lorenzen's structural rules in favor of a more liberal framework (e.g. no strict alternation of players).

Now, assume that **P** has a winning strategy for some dialogue. Then anything implicit in **P**'s original formula is also implicit in the set of premises initially granted by **O**—in a sense, anyone committed to the latter is tacitly committed to the former. This is in harmony with Brandom's logical expressivism: logically complex sentences make explicit proprieties of scorekeeping. In this case they codify what follows from what in an underlying practice of deontic scorekeeping.

6   Conclusion

At present, we regard the question as to whether dialogue games can be consistently interpreted as deontic scorekeeping practices as open. As indicated above, an answer may lead to a significant departure from Lorenzen's original dialogues, and a number of technical problems remain to be solved.

Finally, what about Hodges' charge of circularity? Are scorekeeping practices intelligible without appealing to logical concepts? Can one make sense of individuals engaged in GOGAR unless one ascribes to them genuinely logical motivations to begin with? It is our belief that if dialogue games represent a form of scorekeeping, we can defer these questions to Brandom. His distinction between what is *implicit* in a practice and what can be made *explicit* by means of logic might be just the kind of regress-stopper Hodges thinks is missing. In any case, if Brandom's work can be brought to bear on these foundational debates, this will open up a new and exciting perspective on questions concerning the nature of logic.

References

Brandom, R. B. (1994). *Making it explicit.* Cambridge: Harvard University Press.

Brandom, R. B. (2000). *Articulating reasons.* Cambridge: Harvard University Press.

Brandom, R. B. (2008). *Between saying and doing.* Oxford: Oxford University Press.

Felscher, W. (1986). Dialogues as a foundation for intuitionistic logic. *Handbook of Philosophical Logic*, *3*, 341–372.

Fermüller, C. G. (2008). Dialogue games for many-valued logics—an overview. *Studia Logica*, *90*(1), 43–68.

Fermüller, C. G. (2010). Some critical remarks on incompatibility semantics. In M. Peliš (Ed.), *Logica Yearbook 2009* (pp. 81–95). College Publications.

Fodor, J., & Lepore, E. (2007). Brandom beleaguered. *Philosophy and Phenomenological Research*, *74*(3), 677–691.

Gelfond, M., & Lifschitz, V. (1988). The stable model semantics for logic programming. In *Proceedings of the fifth international conference on logic programming* (pp. 1070–1080).

Giles, R. (1974). A non-classical logic for physics. *Studia Logica*, *33*(4), 397–415.

Hodges, W. (2001). Dialogue foundations: A sceptical look. In *The aristotelian society supplementary volume* (pp. 17–32).

Lorenz, K. (1968). Dialogspiele als semantische Grundlage von Logikkalkülen. *Archive for Mathematical Logic*, *11*(3), 73–100.

Lorenzen, P. (1960). Logik und Agon. *Atti Congr. Internat. di Filosofia*, *4*, 187–194.

Rückert, H. (2001). Why dialogical logic? In H. Wansing (Ed.), *Essays on non-classical logic* (pp. 165–185). World Scientific.

Friedrich Slivovsky
Institut für Computersprachen TU Wien
Favoritenstraße 9-11
Vienna, Austria
e-mail: `fs@logic.at`

What is Wrong with the Tarskian Theory of Truth?

Shawn Standefer*

Abstract

We present a characterization of the Tarskian theory of truth and then present the four major types of objection to it: metalanguage objections, fragmentation objections, expressibility objections, and self-reference objections. We present and discuss different versions of these objections. We argue that only two specific forms of the objections supply good reasons to reject the Tarskian theory.

In many versions of the Tarskian theory of truth, the grammar of the truth predicate bans circular predications. Many philosophers want a *non-Tarskian* theory of truth, which allows for unrestricted ascriptions of truth. The move to a non-Tarskian theory is supposed to surmount objections rooted in the structure of the Tarskian theory and its restricted truth predicates. Objections to Tarskian theories of truth motivate recent philosophical and formal work on truth. Not all of the objections are strong. Consequently, one is left wondering what, precisely, is wrong with the Tarskian theory of truth. We will attempt to answer this question.

In the literature, we find four main types of objections that supply reasons to abandon the Tarskian theory: metalanguage objections, fragmentation objections, expressibility objections, and the self-reference objection. The *metalanguage objection* (§2) tries to create problems for the Tarskian theory on the basis of its distinction between

---

*I have benefitted greatly from discussions of these issues with many people, including Colin Caret, Ole Hjortland, Dave Ripley, the FLC group at Arche, James Shaw, Nuel Belnap and Anil Gupta. I wish to thank the Wesley Salmon Fund for paying for my trip to Logica.

object language and metalanguage. The *fragmentation objection* (§3) focuses on the philosophical problems created by breaking truth into many distinct predicates. The *expressibility objection* (§4) attempts to create problems for the Tarskian theory on the basis of notions that are rendered ineffable by the theory. The *self-reference objection* (§5) highlights problems the Tarskian theory has with apparently unproblematic self-referential truth attributions. Distinct versions of each of these objections are found in the literature. We will discuss the best of each. We argue that only two of the versions of the objections provide good reasons to reject the Tarskian theory: one version of the fragmentation objection and the self-reference objection. We indicate the conclusions that should be drawn from this discussion (§6).

Before presenting the objections to the Tarskian theory, we will make some distinctions and give a brief characterization of the theory (§1).

1   A characterization of the Tarskian theory

At the outset, we should distinguish between *Tarski's theory of truth* and the *broadly Tarskian* theory of truth. By the former, we mean the specific form of the theory of truth presented in (Tarski, 1944) and (Tarski, 1983), which includes Tarski's philosophical views concerning truth. By a broadly Tarskian theory of truth, we mean theories that follow Tarski, either in placing a grammatical restriction on the truth predicate and including an infinite hierarchy of truth predicates or mirroring the hierarchy in the semantics. We will focus on the broadly Tarskian views, rather than Tarski's own view.

We can distinguish two kinds of Tarskian approaches to truth. The first is the *orthodox* approach, which follows the formal details of Tarski's work. Proponents of this approach include Quine and Davidson. The second kind is the *contextualist* approach, which combines a contextual index shift with a Tarskian hierarchy. Proponents of this view include Parsons and Burge.[1] Much of the discussion will apply to both kinds of Tarskians, although our focus is on the orthodox Tarskians.

Tarski's theory of truth included a philosophical commitment that is important for our purposes. Tarski doubted that his methods were

---

[1]See (Parsons, 1974) and (Burge, 1979), respectively.

applicable to natural languages because he thought that no consistent theory of truth could be given for natural languages. Many proponents of Tarskian theories of truth part with Tarski and aim to give such a theory. This difference is crucial to many objections to the Tarskian theory of truth, as we will see.

The central features of the Tarskian theory are its restricted truth predicates and its use of multiple truth predicates, in the case of the broad theory, which we explain shortly. On some versions, the truth predicates are not self-applicable at all. There are sophisticated accounts that can incorporate some amount of self-reference at the syntactic level, but they incorporate the hierarchical aspects into the semantics. One way to eliminate self-reference is for the truth predicate to be in one language, the metalanguage, and only apply to sentences of the language for which it is a truth predicate, the object language, which does not contain a truth predicate for itself. On some ways of proceeding, there is a sequence of languages in addition to a sequence of truth predicates. Some philosophers, such as Quine, advocate using just a single language with a complex grammar. The standard picture, however, is that there are many truth predicates, which reside in sequences of distinct languages.

It will be useful in discussing some of the objections to distinguish two kinds of Tarskian theory, narrow and broad. A *narrow Tarskian theory* is the theory of a particular language, such as the theory $\mathbb{T}(\mathscr{L}_0)$ which is a theory of truth for the language $\mathscr{L}_0$. The theory $\mathbb{T}(\mathscr{L}_0)$ is in the language $\mathscr{L}_1$ and contains a truth predicate T_0 that applies only to sentences of $\mathscr{L}_0$, or their translations. The *broad Tarskian theory* is the combination[2] of all the theories of the hierarchy of languages.[3]

We will now proceed to the first major type of objection, the metalanguage objection.

2 Metalanguages

The first objection that we will consider is the metalanguage objection. The abstract version of this objection says that a theory of truth that requires a metalanguage stronger than the object language

---

[2]We use the term "combination" here to leave open how the theories are merged. None of the objections turn on this detail.

[3]In his writings about truth, Tarski never appealed to the broad theory. Only philosophers after Tarski have done so.

in which to state the theory is inadequate as a theory of truth. For example, suppose we have a Tarskian theory of truth $\mathbb{T}(\mathscr{L}_0)$ for the language $\mathscr{L}_0$, in which vicious self-reference is possible, and that $\mathbb{T}(\mathscr{L}_0)$ is in the language $\mathscr{L}_1$. In order to give a Tarskian theory $\mathbb{T}(\mathscr{L}_1)$ for $\mathscr{L}_1$, one must use a distinct, stronger metalanguage, $\mathscr{L}_2$. The metalanguage $\mathscr{L}_{n+1}$ contains a truth predicate for the object language $\mathscr{L}_n$. This process can be iterated to produce a hierarchy of metalanguages. The objection concludes that the Tarskian theory of truth is inadequate because it is committed to there being such a hierarchy. As an example, Mares describes the situation as follows.

> Consider for example, Tarski's theory of truth. This view commits us to holding that an infinite hierarchy of metalanguages is needed to provide a complete theory of truth. This commitment is a great cost indeed. [Footnote:] Although only one metalanguage is needed to give a theory of truth for a given object language, since we are committed to there being a coherent notion of truth for the metalanguage as well, it requires a metalanguage, and so on. (Mares, 2004, p. 270)

This quotation is representative of a dissatisfaction with the Tarskian theory of truth, with similar sentiments expressed by others.

No single narrow Tarskian theory captures the concept of truth for the whole hierarchy, and the process of generating the content of the broad theory does not culminate in a fixed point. Indeed, the Tarskian theorist can reasonably think that Mares is wrong to say that the theory aims at a complete theory of truth at all. There is no such thing by Tarskian lights, just increasingly richer theories. The completeness aspect is not the heart of the objection though, so we will focus on the commitments of the theory.

The criticism presented by Mares can be interpreted in two ways. The first way is that adopting a Tarskian theory of truth commits us to the existence of an infinite hierarchy of metalanguages and that commitment is by itself reason enough to think the Tarskian theory inadequate.

This is a poor objection to the Tarskian theory because it does not depend on the theory of truth at all. If we take languages to be abstract objects, sets of strings of symbols and interpretations of those, then the infinite hierarchy is not the fault of the Tarskian theory. The languages exist by set existence principles. Even if one does not adopt a Tarskian theory, there will still be infinite hierarchies of Tarskian theories. Problems stemming from language existence are not the fault of the Tarskian theory.

This interpretation does not make the criticism compelling, so we will turn to the other. Mares says that the infinite number of metalanguages is a great cost. We take it that he has the following sort of argument in mind.

M1 If a theory of truth requires an infinite number of increasingly stronger metalanguages to give a complete description of truth, then it is unacceptable as a theory of truth for natural language.[4]

M2 The Tarskian theory of truth requires an infinite number of increasingly stronger metalanguages to give a complete description of truth.

M3 So, the Tarskian theory of truth is unacceptable as a theory of truth for natural language.

We have mentioned that a Tarskian would have issues with (M2). We want to focus on the first premiss. We can argue for (M1) using the following two supplementary premises.

S1 An adequate theory of truth must be applicable to natural languages.

S2 Natural languages have no strictly stronger metalanguages.

Supplementary premiss (S1) lets us focus on English (or another natural language).[5] The goal of many theorists is to give a theory of truth for natural language. The first premiss is largely accepted by Tarskians but not by Tarski. Supplementary premiss (S2) is more contentious and it generates the problems for the Tarskian theory. It does not even require an infinite hierarchy of metalanguages to cause problems for the Tarskian, because a single such metalanguage would provide the basis of an objection.

---

[4]It is an interesting question what is meant by a *complete* theory of truth. Unfortunately, we cannot discuss this here.

[5]We will drop the parenthetical "(or another natural language)" in what follows and talk about English alone. This should be understood with the additional content of the parenthetical where appropriate.

The argument for (M1) runs as follows. According to (S1), an adequate theory of truth must be applicable to natural languages, such as English. By (S2), there is no strictly stronger metalanguage for English. A theory of truth that requires a strictly stronger metalanguage to provide a complete description of truth cannot be used as a theory of truth for English. Therefore, such a theory of truth is inadequate as a theory of truth for English. Now that we have presented the argument, we will sketch a motivating intuition for (S2) before presenting an objection to it.

The idea behind (S2) seems to be this. Speakers of, say, English can adopt a fragment of English as a metalanguage to talk about truth for another proper fragment of English. Talking about truth for the whole language, however, would require using some stronger language that goes beyond one's own. That would require that the speaker use a language that is distinct from her own language, where one's own language is exhaustive, meaning that if a speaker does use some vocabulary, it is part of her own language.[6] This, however, is impossible. Thus, proponents of (S2) conclude that English has no strictly stronger metalanguages.

The problem we find with (S2) is that there is not good reason to think it true. It uses the notion of expressive power that does not straightforwardly apply to natural languages in a way that makes this come out true. We speculate that some hazy version of Tarski's claim that natural languages are universal is motivating this. Universality is a murky notion itself, so it is not going to help in bolstering (S2) without first receiving clarification.

We conclude that the metalanguage objections do not provide good reason to abandon the Tarskian theory. The metalanguage objection seems to be aimed at problems more properly described as worries about fragmentation, which we will now discuss.

3 Fragmentation, unity, and hierarchies

The second major objection is that the Tarskian theory fragments a unified concept.[7] Our ordinary notion of truth is a unified one, neither schematic nor ambiguous. There is just one intuitive concept of

---

[6]This idea seems to be needed to make sense of the alleged impossibility.

[7]For this discussion, I make the simplifying assumption that the logical behavior of predicates is the same as that of the concepts they express. I will,

truth.[8] The Tarskian theory breaks this concept into many concepts, each of which can be viewed as part of ours. Viewed abstractly, the fragmentation objection says that fragmenting truth in the manner of the Tarskian theory entails negative consequences, which make the Tarskian theory inadequate. The specific negative consequences depend on the concrete form of the objection. There are three forms that the fragmentation objection takes: the philosophical fragmentation objection (§3.1), the scientific fragmentation objection (§3.2), and the simple unity objection (§3.3).

3.1 Philosophical fragmentation

The first form of fragmentation objection is the *philosophical fragmentation objection*. Truth is an important concept in philosophy, and concepts analyzed in terms of truth will be similarly fragmented. As an example, the fragmentation of truth breaks the concept of knowledge up into a hierarchy of knowledge predicates, each of which uses a different truth predicate. Using a knowledge predicate with a sufficiently high index, many things of interest could be ratified as knowledge. A unified, general account, however, will not be possible while using the Tarskian theory of truth. The situation would be the same with any such fragmented concept. This is because we would be precluded by the structure of the Tarskian theory from saying some general things about the fragmented concepts. The factivity of knowledge is an example. A natural rendering of factivity is this: $\forall x(Knows(x) \rightarrow T(x))$. Using a Tarskian truth predicate, we would need to put it schematically, $\forall x(Knows_n(x) \rightarrow T_n(x))$, with the subscripted '$n$'s as schematic letters for indices. Insofar as philosophers aim to give unified, general theories of some concept that depends on the concept of truth, their goals are incompatible with a Tarskian approach to truth. Therefore, the objection concludes, any such philosopher should reject the Tarskian theory of truth.

---

consequently, pass back and forth between talking about predicates and concepts. The Tarskian could respond by saying that, while her theory of truth has many truth predicates, there is only one concept of truth it is getting at. This does not seem to us to be a promising line of response. We cannot respond to it here.

[8]We will suppress the qualification needed on the intuitive, semantic concept of truth. A qualification is needed, since there are expressions such as "true friend" that indicate other concepts that are expressed by the word 'true'. This was pointed out to us by Nuel Belnap.

3.2 Scientific fragmentation

The second form of fragmentation objection is the *scientific fragmentation objection*. This is similar to the previous form, except it focuses on the results of the fragmentation for scientific inquiry. McGee puts the point as follows.

> If we adopt [Tarski's theory of truth], we shall find that within the object language we are unable even to describe human thought and action. ... [Within the metalanguage] we can describe intentional human activities that are directed toward inanimate objects, but thought about thought and talk about talk will remain indescribable and inexplicable. Thus, if we accept the limitations imposed by Tarski's proposal for avoiding antimonies, we forfeit one of the highest aspirations of the human spirit, the aspiration to self-understanding. (McGee, 1991, pp. 78–79)

Much can be said about thought and speech, even thought and speech about thought and speech, but a unified, general account will not be possible. Scientific investigations of intentional human behavior will, presumably, involve semantic concepts such as truth, yet would be unable to make fully general claims about agents due to the involvement of truth. The fragmentation of truth prevents there from being unified theories of psychology and linguistics. The objection does not need one to subscribe to a global unity of science thesis, that is, the claim that there is a single, unified theory of all the sciences, but rather the localized claim that there are unified theories of the individual sciences. The Tarskian theory of truth poses a challenge to this weakened form, if developed theories of psychology and linguistics use semantic concepts. This is troubling because there should not be *a priori* objections from the theory of truth to the possibility of unified linguistic and psychological theories. Therefore, the objection concludes, theories of truth that fragment the truth predicate should be rejected.

3.3 Simple unity

The third form of fragmentation objection is the *simple unity objection*. This form says that there is a single, unified concept of truth. The fragmentation of truth is objectionable because a theory of truth should aim at explaining the unified truth predicate that is employed in colloquial language. A theory that breaks the concept of truth into a hierarchy is talking about a distinct concept. According to this objection, a theory that offers a hierarchical concept of truth is, at best, offering a replacement for the intuitive one, but that does not illuminate the way the intuitive concept works. The objection concludes that any theory of truth that fragments the truth predicate should be rejected as an inadequate account of the intuitive concept of truth.

3.4 Evaluation of the fragmentation objection

The fragmentation objection has three forms. The force of the first form, philosophical fragmentation, depends on what project and concepts would be barred. Knowledge was offered as an example, but this will be undermined if there are alternative analyses that do not appeal to a truth predicate in the way indicated. Standard presentations of the factivity of knowledge use either schematic propositional letters or propositional quantifiers, so it is not much of a cost to use a schematic form of factivity. One could accept that the fragmentation is unfortunate but not the basis of a decisive objection. In the absence of a compelling example, this objection is not decisive.

We now turn to the second form, scientific fragmentation. We want to put pressure on something on which it depends: the role of generality in science. To asses the objection, we must distinguish two sorts of generality: the $\exists\forall$ sort and the $\forall\exists$ sort. The $\exists\forall$ sort of generality for science says that there is one theory that correctly describes all types of phenomena. The $\forall\exists$ sort says that all types of phenomena are accurately described by some theory. The question, then, is whether science demands $\exists\forall$ generality or whether $\forall\exists$ generality suffices. Contemporary physics seems to be a case in which we have only $\forall\exists$ generality. General relativity correctly describes the very large and quantum mechanics the very small, but the two do not combine well. The goal is to get $\exists\forall$ generality in physics, but one would expect that theoretical physics would not be viewed as a failure should this not be available. There is no need to demand more of semantics.

This line of response provides adequate reason not to reject the Tarskian theory on the basis of the scientific fragmentation objection. We will now discuss the simple unity objection.

The Tarskian has a response to the simple unity objection that

we should examine.[9] While the intuitive concept of truth is not fragmented, there are examples of unified concepts being broken apart
in analysis. The concept of jade, for example, is a single intuitive
concept that is split into multiple concepts in more precise theoretical treatments. This provides some evidence that breaking a unified
concept apart in philosophical analysis is not always bad.

The simple unity objector has a response explaining how truth is
different than other intuitively unified concepts that are fragmented
in theoretical reconstructions. Splitting a concept is not in itself a
problem. If we are engaged in a descriptive project, there could be
facts about our usage or the objects that support splitting concepts.
Jade is like this, which is why splitting the concept of jade is unobjectionable. In the case of truth, no such facts are readily available. The
main option is to appeal to facts about linguistic practice to support
the fragmentation, but, *prima facie*, this will not justify fragmenting
truth in the Tarskian way. Thus, truth is different from intuitively
unified concepts, such as jade.

The responses given in this section to the philosophical fragmentation and the scientific fragmentation versions of the fragmentation
objection are adequate responses. The simple unity objection, however, remains a problem for the Tarskian.

4   Expressing the theory

The third major form of objection to the Tarskian theory of truth
is the *expressibility objection*. The abstract form of the expressibility
objection notes that there is some concept that we intuitively understand, and that is, further, within the area of application of the theory.
Because of this, it should be formalizable in our theory. The theory
is barred from expressing this concept because adding the concept to
the theory results in contradiction. The theory is, thus, flawed, so
there is reason to reject the theory.

The main version of this objection focuses on the hierarchy of
languages and truth predicates in the broad Tarskian theory. This is
most naturally directed at the orthodox Tarskian, although a variant
could be produced for the contextualist. In constructing the broad
theory, we employ the concept of being true at some level, and, further,

---

[9]The suggestion for this line of thought came from James Shaw.

it seems reasonable to say that a sentence is true at some level. True at some level is not something that can be formulated generally from within the Tarskian hierarchy, because this quantifies into the index position on the truth predicates, which is not standardly open to quantification. The concept of truth at some level, however, intuitively makes sense. This is some indication that the proper logical form of the Tarskian truth predicates is a binary relation that makes the index parameter i available for quantification: $T(x, i)$. The quantified form, $\exists y T(x, y)$, will lead to paradox. The objection says that the Tarskian theory blocks this paradox only through *ad hoc* restrictions on the grammar of the predicate. The objection concludes that the Tarskian theory is thus inadequate as a theory of truth for natural language.

This objection provides the Tarskian good reason not to adopt the broad theory. The Tarskian does not seem to be committed to using or defending the broad theory. She can claim that the broad theory is merely a metaphor used in explaining the method of constructing new narrow theories. She need only defend the coherence of this method. This argument does not provide good reason to reject a Tarskian approach that uses only narrow theories, as Tarski did, for example.

5   Self-reference

The last major objection, presented in (Kripke, 1975), points out that in colloquial language we frequently attribute truth in a circular way that resists Tarskian stratification. This is exemplified by Kripke's Dean-Nixon case, in which Dean says that everything Nixon says is true and Nixon says that everything Dean says is false, and that is all that either says. The truth attributions of the two utterances cannot be unwound as required by the Tarskian theory. This objection can be extended from circular ascriptions to intuitive reasoning with circular ascriptions.[10] Since the Tarskian theory of truth is unable to account for these cases, it is inadequate as a theory of truth for natural language.

There are, additionally, many cases where a sentence will be para-doxical or circular depending on the empirical facts, as the example shows. Kripke says that "an adequate theory of truth must allow our statements involving the notion of truth to be *risky*: they risk being

---

[10]See (Gupta, 1984, pp. 210–211) for examples.

paradoxical if the empirical facts are extremely (and unexpectedly) unfavorable."[11] A theory of truth for a natural language must not be incompatible with certain unfortunate combinations of facts.

Kripke thinks that we need to have a self-referential truth predicate. We think that Kripke's assessment is correct here and that this objection to the Tarskian theory is a strong one.

6   Conclusions

The Tarskian theory of truth comes with costs, such as restricted truth predicates, and merely apparent costs, such as employing a distinction between metalanguage and object language. The theory has its virtues, such as unrestricted use of classical logic. We have presented several objections to the Tarskian theory of truth that have, in some form, motivated recent work on non-Tarskian theories of truth. We have argued that many of these objections, once clarified, have turned out to be poor and do not provide good reason to reject the Tarskian theory of truth. There are, however, reasons to want a non-Tarskian theory, which brings us back to the titular question.

There are two strong reasons not to be satisfied with the Tarskian theory of truth. The first is Kripke's point that there are many sentences of colloquial language that are circular, and are so depending on the facts. A satisfying descriptive theory of truth should be compatible with this. The second is the simple unity objection. The intuitive concept of truth is unified. If we are concerned with explaining that concept, we should provide a theory for a non-fragmented concept. Once we adopt a non-Tarskian theory of truth, we should ask how well that theory responds to these objections.

The two reasons for rejecting the Tarskian theory depend on the claim that an adequate theory of truth must be applicable to natural languages.[12] Insofar as the content and status of the claim that a theory of truth should be a theory of truth for natural language is unclear, these reasons will lack force.

---

[11](Kripke, 1975, p. 692)

[12]As noted in §1, this was not a goal of Tarski's.

References

Burge, T. (1979). Semantical paradox. *Journal of Philosophy*, *76*(4), 168–198.

Gupta, A. (1984). Truth and paradox. In R. L. Martin (Ed.), *Recent essays on truth and the liar paradox*. Oxford University Press.

Kripke, S. (1975). Outline of a theory of truth. *Journal of Philosophy*, *72*, 690–716.

Mares, E. (2004). Semantic dialetheism. In J. Beall & G. Priest (Eds.), *The law of non-contradiction* (pp. 264–275). Oxford.

McGee, V. (1991). *Truth, vagueness, and paradox: An essay on the logic of truth*. Cambridge.

Parsons, C. (1974). The Liar Paradox. *Journal of Philosophical Logic*, *3*(4), 381–412.

Tarski, A. (1944). The semantic conception of truth: And the foundations of semantics. *Philosophy and Phenomenological Research*, *4*(3), 341–376.

Tarski, A. (1983). The concept of truth in formal languages. In J. Corcoran (Ed.), *Logic, semantics, metamathematics*. Hackett.

Shawn Standefer
Philosophy Department
University of Pittsburgh
Pittsburgh, PA 15260 USA
e-mail: `standefer@gmail.com`

Infinite Natural Numbers: an Unwanted Phenomenon, or a Useful Concept?

Vítězslav Švejdar*

Abstract

We consider non-standard models of Peano arithmetic and non-standard numbers in set theory, showing that not only they appear rather naturally, but also have interesting methodological consequences and even practical applications. We also show that the Czech logical school, namely Petr Vopěnka, considerably contributed to this area.

1 Peano arithmetic and its models

Peano arithmetic PA was invented as an axiomatic theory of natural numbers (non-negative integers 0, 1, 2, ...) with addition and multiplication as designated operations. Its language (arithmetical language) consists of the symbols $+$ and $\cdot$ for these two operations, the symbol 0 for the number zero and the symbol S for the successor function (addition of one). The language $\{+, \cdot, 0, S\}$ could be replaced with the language $\{+, \cdot, 0, 1\}$ having a constant for the number one: with the constant 1 at hand one can define $S(x)$ as $x + 1$, and with the symbol S one can define 1 as $S(0)$. However, we stick with the language $\{+, \cdot, 0, S\}$, since it is used in traditional sources like (Tarski, Mostowski, & Robinson, 1953). The closed terms $S(0)$, $S(S(0))$, ... represent the numbers 1, 2, ... in the arithmetical language. We write $\bar{n}$ for the numeral $S(S(.. S(0)..))$ with n occurrences of the symbol S. So $\bar{0}$ and 0 are the same terms. The *standard model* of Peano

*This work is a part of the research plan MSM 0021620839 that is financed by the Ministry of Education of the Czech Republic.

arithmetic is the usual structure of natural numbers, i.e. the struc-
ture $\mathbb{N} = \langle N, +^N, \cdot^N, 0^N, s \rangle$, where s is the function $a \mapsto a + 1$ and the
remaining symbols have the obvious meaning. Since it is not difficult
to distinguish symbols from their realizations, we will often omit the
superscripts when dealing with $+^N$, $\cdot^N$, and 0^N.

The axioms of Peano arithmetic are (e.g. in Tarski et al., 1953) for-
mulated as the induction scheme $\varphi(0) \& \forall x(\varphi(x) \to \varphi(S(x))) \to \forall x \varphi(x)$
together with seven simple axioms Q1–Q7 that PA shares with Robin-
son arithmetic Q (where $\forall x(x + 0 = x)$, the axiom Q4, is a sample).
The induction scheme stipulates that if the number zero has a property expressible in the arithmetical language and if it is the case that
whenever a has that property then also $a + 1$ has that property, then
all numbers have that property. In some sources the language of PA
contains also the symbols $\leq$ and $<$ for non-strict and strict order.
However, we can speak about an order of numbers anyway. One can
define that $x \leq y$ iff $\exists v(v + x = y)$ i.e. iff y is a result of an addition
of x and some other number, and then one can define that $x < y$ iff
$x \leq y$ and $x \neq y$ (or equivalently, iff $\exists v(S(v) + x = y)$). With the order
at hand, one can formulate the *least number principle* saying that if
there exist numbers having a property expressible in the arithmetical
language, then there exists a least number having that property. It
is not difficult to verify that the least number principle is basically
equivalent to the induction scheme. More precisely, the two schemes
are equivalent over Robinson arithmetic equipped with an additional
axiom $\forall x(x < S(x))$.

Validity of the least number principle in a model means that every
set described by an arithmetical formula (every *definable set*), if it
is non-empty, has a least element. Thus the least number principle
is automatically valid in the standard model $\mathbb{N}$ since the standard
model is well-ordered, which means that *every* non-empty set has a
least element. However, an extremely interesting observation is that
if the model is not well-ordered, it is still possible that all definable
non-empty sets have a least element; hence models of PA different
from $\mathbb{N}$ may exist.

Thus one can define that a model $\mathbb{M}$ of PA is *non-standard* if
(a) contains an element which is not accessible from zero by a finite
number of steps of the successor function, or (b) if its order defined
by $x < y$ iff $\exists v(S(v) + x = y)$ is not a well-order. One can easily
check that the conditions (a) and (b) are equivalent. Indeed, an order

in which every element except the very least one has a predecessor and for each element a there exists only a finite number of elements smaller than a must be a well-order (i.e. a linear well-founded order). On the other hand, if there are elements not accessible from zero by a finite number of steps, these constitute a non-empty set not having a least element.

In the standard model, every element is a value of some of the numerals $\bar{0}, \bar{1}, \bar{2}, \ldots$ In a non-standard model (if such exist), there are elements greater than the values of all numerals—and of all closed terms. These elements are called *non-standard*, while the other elements are *standard*. Standard elements precede the non-standard ones.

Non-standard models are not just a logical possibility, but a reality. Nowadays a simple way to prove their existence is extending the language of PA by a constant c and considering an auxiliary theory T in this language, whose axioms are the axioms of PA together with infinitely many additional axioms $\bar{0} < c, \bar{1} < c, \bar{2} < c, \ldots$ Any finite set F of axioms such that $F \subseteq \mathrm{PA} \cup \{\, \bar{n} < c \,;\, n \in \mathrm{N} \,\}$ has a model: it can be obtained by taking the standard model $\mathbb{N}$ and realizing the constant c by its sufficiently big element. Then the compactness theorem says that T has a model $\mathbb{M}$. In $\mathbb{M}$, the constant c is realized by a non-standard element. The reduct of $\mathbb{M}$ to the arithmetical language, obtained by omitting the constant c but not changing the domain and realizations of arithmetical operations, is a non-standard model of PA.

At first sight, the non-standard models look like an unwanted phenomenon that either shows that some important axioms are missing in the axiomatic system of PA, or demonstrates some deficiency of the classical first order logic. However, no additional axioms can help to prevent this phenomenon, since it is the case that all consistent extensions of PA have non-standard models. And instead of correcting the first order logic, I would opt for thinking about an expressive power of formalized languages and about using non-standard models in mathematical practice. As to the expressive power, consider for example the following properties and conditions for natural number: (a) x is less than y, (b) y is a power of 2, (c) y is the x-th power of 2, i.e. $y = 2^x$, and (d) x is accessible from 0 by finite number of steps of the successor function. With some effort and probably first having developed some coding of sequences, one can show that (c) is expressible by an arithmetical formula. With much less effort (in-

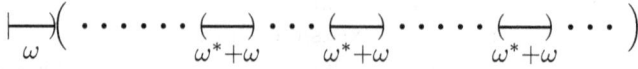

Figure 1: A non-standard model of PA

deed, this is a nice homework) one can show that (b) is expressible, and we have already seen that (a) is expressible as well. Non-standard models of PA show that the property in (d) is *not* expressible in the arithmetical language, because otherwise one could show, using the inductions axiom, that *all* numbers have that property. And similarly as with extending the axiom set, extending the language is of no help: theories with a language containing that of PA also have non-standard models.

2   The order structure of a non-standard model

Let $\mathbb{M}$ be a non-standard model of PA and let a be its non-standard element. From the fact that all theorems of PA are valid in $\mathbb{M}$ we can conclude that a is by far not the only non-standard element of $\mathbb{M}$. For example, every number $x \neq 0$ has a predecessor, i.e. a number y such that $S(y) = x$ and $y < x$ and there are no other numbers between y and x. So our element a of $\mathbb{M}$ has a predecessor that can reasonably be denoted $a - \bar{1}$ even if there is no symbol for subtraction in the arithmetical language. This $a - \bar{1}$ must be non-standard. We can continue and consider numbers $a - \bar{2}$, $a - \bar{3}$, ...; all of these must be non-standard. Going upwards, we can consider numbers $a + \bar{1} < a + \bar{2} < a + \bar{3} < \ldots$ So we see that our non-standard number a is surrounded by a cluster $[a]$ of infinitely many other non-standard numbers whose distance from a is finite (standard). The order type of the set $[a]$ is that of integers and can schematically be denoted $\omega^* + \omega$ where ω is the order type of natural numbers, ω^* is the reversed order of natural numbers, and $+$ denotes the disjoint sum of the two structures where the elements of ω^* precede all elements of ω. Besides this cluster $[a]$ there is the initial cluster $[0]$ of all standard numbers; the order type of this cluster is ω.

Still, $[0]$ and $[a]$ cannot be the only clusters in $\mathbb{M}$. The cluster $[\bar{2} \cdot a]$ is different from $[a]$ since the distance between a and $\bar{2} \cdot a$ is a, a non-standard number. Similarly, the cluster $[a \cdot a]$ is different from the pairwise different (and disjoint) clusters $[a]$, $[\overline{2} \cdot a]$, $[\overline{3} \cdot a]$, ... There is no greatest cluster. And we can still continue and show that the clusters are *densely* ordered. Schematically, the order structure of $\mathbb{M}$ is $\omega + (\omega^\star + \omega) \cdot \xi$, where ξ is a linear dense order without endpoints and the multiplication symbol $\cdot$ denotes the operation of replacing each element of ξ by the structure $\omega^\star + \omega$, see Fig. 1. If the model $\mathbb{M}$ is countable then its order structure is $\omega + (\omega^\star + \omega) \cdot \eta$ where η is the uniquely determined countable linear order without endpoints (i.e. the order of rationals).

It must be emphasized that what we do in this section is *not a construction* of a non-standard model of PA, but a reasoning about its order once its existence has been proved. We have presented the existence of a non-standard model of PA as a consequence of the compactness theorem, and then determined its structure by making use of the knowledge that some sentences, as theorems of PA, must be valid in it.

A second thing that has to be emphasized is that a one-one function from one structure on another that preserves order and successor does not necessarily preserve addition and multiplication. So two models that are order isomorphic are not necessarily isomorphic as structures for the arithmetical language. Indeed, non-isomorphic countable models of PA—necessarily having the same order structure—do exist.

3 Some history

It was Thoralf Skolem who proved the existence of non-standard model of PA in (Skolem, 1934). The earlier 1920 and 1922 papers of Skolem contain a proof of Löwenheim-Skolem theorem, saying that if a theory with at most countable language has an infinite model then it also has a countable model. A then surprising consequence of this theorem was the existence of countable models of set theory; this fact is known as *Skolem paradox*. It is not clear (to me) whether Skolem was then aware of the stronger variant of that theorem, saying that if a theory with at most countable language has an infinite model then it has models of all infinite cardinalities. This stronger variant of Löwenheim-Skolem theorem entails that PA has uncountable—and hence necessarily non-standard—models.

From Gödel 1st incompleteness theorem, published in 1931, we know that PA is incomplete. From that (and from the completeness theorem published also by Gödel in 1930 but perhaps known to Skolem even before 1930) it is clear that PA has models that differ in validity of some sentences. If two models differ in validity of some sentences then at least one of them is non-standard. This proof of existence of non-standard models is much more involved than the proof via the compactness theorem (because it in fact contains some considerations about recursive functions), but was available some time before Skolem's 1934 paper. So one could ask why, in the light of Gödel's paper and Skolem's earlier papers, the Skolem's 1934 paper is so important. The answer is that Skolem's primary interest then were not models that differ in validity of some sentences, but models that are non-isomorphic while not being distinguishable by validity of some sentences. Skolem thus invented the notion of *elementary equivalence* of models; by doing that he became a pioneer of model theory. It is important to remark that the 1934 paper contains not just a proof of existence, but a direct construction of a non-standard model.

In the Czech logical environment, it was Ladislav Svante Rieger (1916–1963) who knew about the existence and was familiar with a construction of non-standard models. Rieger's interest was algebraic logic, probably in the style of Rasiowa and Sikorski, and was the inventor of Rieger-Nishimura lattice, a beautiful structure of infinitely many intuitionistically non-equivalent formulas built up from one propositional atom only, see (Rieger, 1949). Rieger was initially an official thesis advisor of a brilliant Czech logician Petr Hájek. However, Rieger died soon, well before Hájek wrote his thesis, and Hájek never fails to mention another brilliant Czech logician *Petr Vopěnka* (born 1935) as his teacher. Vopěnka was a student of Eduard Čech, the inventor of Čech-Stone compactification in topology.

It is unclear and probably unknown whether Rieger's construction of a non-standard model of PA was that of Skolem, or his own. It however is known that it was rather complicated. A feasible construction of a non-standard model of PA was given by Vopěnka around 1960. Vopěnka uses ultraproduct and he invented that construction independently of A. Robinson (who uses ultraproduct as well).

The notion of a non-standard model can easily be extended to Zermelo-Fraenkel set theory ZF or to Gödel-Bernays set theory GB. A model $\mathbb{M}$ of (some) set theory is non-standard if it contains an ordinal α such that $\mathbb{M} \models$ "α is finite", i.e. $\mathbb{M} \models$ "α is less than the first limit ordinal", but looking from outside, the set of all ordinals $\beta < \alpha$ is infinite. The proof of the existence of non-standard models of set theory is basically the same as for PA. It seems that Rieger and Vopěnka preferred thinking about set theory to thinking about PA.

4 Definable cuts

In the following definition, we use a somewhat vague notion of a theory *with natural numbers*. In PA or Q, this notion is trivial since all their individuals are natural numbers. In set theory, the natural numbers are all ordinals less than the first limit ordinal ω. Note that here the meaning of the symbol ω is not the same as when we discussed order types.

Definition 1 Let T be a theory with natural numbers, let x be a variable for natural numbers. A formula $J(x)$ is a *(definable) cut in T* if $T \vdash J(0)$ and $T \vdash \forall x (J(x) \to J(S(x)))$.

Let, for example, T be Robinson arithmetic Q and let $J_1(x)$ be the formula $x + 0 = x$. Then the formula J_1 is a trivial cut because the axiom Q4, saying that $\forall x(x + 0 = x)$, in fact says $\forall x J_1(x)$. If $J_2(x)$ is chosen as the formula $0 + x = x$ then the situation is more interesting since $\forall x J_2(x)$ is known as being unprovable in Q. However, $Q \vdash J_2(0)$ follows from the axiom Q4 and $Q \vdash \forall x (J_2(x) \to J_2(S(x)))$ follows from another axiom Q5, stipulating that $\forall x \forall y (x + S(y) = S(x + y))$. Thus J_2 is a cut in Q.

If a cut J in T is non-trivial, i.e. if $T \nvdash \forall x J(x)$, then there exist models $\mathbb{M}$ of T with elements a such that $\mathbb{M} \nvDash J(a)$. Such an a must necessarily be non-standard in $\mathbb{M}$. There are no non-trivial cuts in PA because they would directly violate induction. If $J(x)$ is a cut in ZF then it follows from the separation axiom that there exists a set A of all natural numbers x such that $\neg J(x)$, viz $A = \{x \in \omega; \neg J(x)\}$. From the definition of cut we know that A has no least element. However, the fact that every non-empty subset of ω has a least element is a theorem of ZF; thus $A = \emptyset$. This argument shows that there are no non-trivial cuts in ZF; we have *full induction* (induction for all existing formulas of its language) in ZF.

We will show an interesting construction, invented in (Vopěnka & Hájek, 1973), of a non-trivial cut in Gödel-Bernays set theory GB.

The construction shows that GB is not a theory with full induction. Later we will discuss some other consequences. Recall that, in GB, the primitive notion is *class*, while a set is defined as a class which is an element of some (other) class. Recall also that GB, as a strong theory, is capable of formalizing logical syntax. That means that inside GB we have the notion of formalized syntactical objects. Out of all syntactical objects, we only need set formulas, i.e. formulas of ZF, and variables in these formulas. We identify formulas and variables with their numerical codes assigned to them by some fixed coding of syntax. Thus we can talk about a formula as being, for example, smaller or greater than a natural number x. When speaking inside GB, formulas are finite objects; when looking at a model of GB from outside, formulas are its natural numbers that can be both standard and nonstandard. An *evaluation of variables* is any function defined on the set of all variables. Thus the domain of an evaluation is the set of all those natural numbers that are (numerical codes of) variables; the values of an evaluation of variables are sets (not proper classes, of course).

Definition 2 (in GB) A relation R between formulas less than x and evaluations of variables is a *truth relation on x* if the following conditions hold.

$$[\varphi \,\&\, \psi, e] \in R \Leftrightarrow [\varphi, e] \in R \text{ and } [\psi, e] \in R \tag{i}$$

whenever $\varphi \,\&\, \psi$ is (and thus both φ and ψ are) less than x; and similarly for other logical connectives.

$$[\forall x \varphi, e] \in R \Leftrightarrow \forall a([\varphi, e(x/a)] \in R) \tag{ii}$$

whenever $\forall x \varphi$ (and thus φ itself) is less than x; and similarly for the other quantifier $\exists$. Here $e(x/a)$ is the evaluation whose value in x is a and the remaining values are the same as the values of e.

$$[x \in y, e] \in R \Leftrightarrow e(x) \in e(y). \tag{iii}$$

In short, a truth relation on x is a relation satisfying the Tarski's truth conditions wherever they are applicable, i.e. whenever the formulas in question are less than x. Note that in the left side of (ii) the quantifier $\forall$ is a formal symbol (part of the formalized syntax), while

	...	e_1	...	...	e_2	...	...	...
		⋮			⋮			
φ_1	...	1	...	...	1	...	...	...
		⋮			⋮			
φ_2	...	0	...	...	1	...	...	...
		⋮			⋮			
$\varphi_1 \mathbin{\&} \varphi_2$	...	0	...	...	1	...	...	...

Figure 2: A truth relation

"$\forall a$" in the right is an abbreviation in our speech about the syntax (a shorthand for "for each set a"). Similarly, "$\in$" in the left of (iii) is a formal symbol (part of the atomic set formula "$x \in y$"), while in the right we say that "$e(x)$, the value assigned to the variable x by the evaluation e, is an element of $e(y)$".

A truth relation is an object like in Fig. 2: a zero-one table where 1 stands for "yes" ($[\varphi_2, e_2] \in R$, for example) and 0 stands for "no". The table has only a finite number of lines (finite in the sense of GB, i.e. standard or non-standard in its model) but a huge number of columns (indeed, the class of all evaluations of variables is a proper class).

One can prove by induction on $y \le x$ that (a) there exists at most one truth relation on x. Also, (b) the empty class is (the only) truth relation on 0. Let a number x be called *occupable* if there exists a truth relation on x. In symbols,

$$\mathrm{Ocp}(x) \Leftrightarrow \exists R (R \text{ is a truth relation on } x).$$

We know from (b) that $\mathrm{Ocp}(0)$, and it is possible to verify (c) that if $\mathrm{Ocp}(x)$ then $\mathrm{Ocp}(\mathrm{S}(x))$. Indeed, let R be a truth relation on x and distinguish the cases whether x is not a formula, is obtained from smaller formula(s) using a logical connective, or is obtained from a smaller formula using quantification. If, for example, x is the disjunction $\varphi \vee \psi$, the relation

$$R' = R \cup \{ [x, e] \; ; \; [\varphi, e] \in R \text{ or } [\psi, e] \in R \},$$

with an additional line for $x = \varphi \mathbin{\&} \psi$, is a truth relation on $\mathrm{S}(x)$.

We see that the formula $\mathrm{Ocp}(x)$ is a cut in GB. If $\{ x \in \omega ; \mathrm{Ocp}(x) \}$ were a class then it would be a set and then it would equal ω: indeed, the fact that if a subset of ω contains 0 and is closed under successor then it equals ω is a theorem of GB (as well as it is a theorem of ZF). The axioms of GB (namely, the comprehension scheme) guarantee that any *normal formula* of GB (one that does not contain quantification of classes) determines a class. However, the formula $\mathrm{Ocp}(x)$ expresses a property of natural numbers which is not normal (as "$\exists R$" in "$\exists R(R$ is a truth relation on $x)$" is a quantification of a proper class).

If all natural numbers are occupable then we could develop a notion of truth for all set formulas, show that all axioms of ZF are true and that deduction rules preserve truth. Then ZF would necessarily be consistent, and also GB would be consistent since GB knows that GB and ZF are equi-consistent. Thus we see that to the first observation, that is is not sure that the formula $\mathrm{Ocp}(x)$ determines a set (a class) because it is not normal, we can add a second observation that a proof that every number is occupable would yield a contradiction with Gödel 2nd incompleteness theorem. So GB $\nvdash \forall x \mathrm{Ocp}(x)$, the formula $\mathrm{Ocp}(x)$ is a non-trivial cut in GB.

5 Some conclusions

The facts that we do not have full induction in GB and that not every formula of GB determines a class show interesting differences between ZF and GB, two theories that otherwise are closely related: GB is conservative over ZF with respect to set formulas.

Definable cuts in GB also show why structures and models in the semantics of first order predicate logic are defined as sets rather than classes. The reason is that the axioms of GB are not strong enough for that generalized definition to work. The (Tarski's) definition of satisfaction, that uses recursion on the structure of formulas, is not quite innocent. Even in the "normal case", where structures and models are sets, the definition needs some axiomatic strength to work. This is also somewhat surprising: the logical semantics that we teach in elementary logic courses is by no means finitistic, it is quite dependent on mathematics, i.e. on some axiomatic theory like ZF.

We see that non-standard natural numbers naturally occur. However, more is true: they can be a useful and applicable tool. We will mention two application. One of them is in logic. R. Solovay in

(Solovay, 1976) used occupable numbers as a method for constructing interpretations in GB, and constructed a set sentence (in fact an arithmetical sentence) φ such that GB, φ is interpretable in GB but ZF, φ is not interpretable in ZF. This unpublished letter answered a question raised by P. Hájek and was an important milestone in the research of interpretability. Before that, the fact that the closely related theories ZF and GB differ in interpretability was proved by P. Hájek.

A second application of non-standard models is *non-standard analysis*. A non-standard model of PA can easily be extended to an ordered field. Then if x is a non-standard number, the number $1/x$ is non-zero but infinitely small. Thus the old Leibniz's idea of *infinitesimals* can be reconstructed and made rigorous using non-standard numbers. As an example, we can give the "reconstructed" definition of continuousness: a function f is *continuous* in a if, for every infinitesimal dx, the value $f(x + dx)$ is infinitely close to $f(x)$, that is, if $|f(x + dx) - f(x)|$ is infinitesimal. In the reconstructed analysis, there is much less quantifiers than in the "modern" (ϵ–δ) analysis.

The idea that infinitesimals can be reconstructed using non-standard natural numbers is A. Robinson's. It was however independently invented by Vopěnka around 1960. In Vopěnka's Alternative Set Theory AST, see (Vopěnka, 1979), there are only two infinite cardinals, but infinitesimals do exist.

References

Rieger, L. S. (1949). On lattice theory of Brouwerian propositional logic. *Acta Facultatis Rerum Naturalium Univ. Carolinae*, *189*, 1–40.

Skolem, T. (1934). Über die Nicht-charakterisierbarkeit der Zahlenreihe mittels endlich oder abzhlbar unendlich vieler Aussagen mit ausschliesslich Zahlenvariablen. *Fundamenta Mathematicae*, *23*, 150–161.

Solovay, R. M. (1976). *Interpretability in set theories*. (Unpublished letter to P. Hájek, Aug. 17, 1976, http://www.cs.cas.cz/~hajek/RSolovayZFGB.pdf)

Tarski, A., Mostowski, A., & Robinson, R. M. (1953). *Undecidable theories*. Amsterdam: North-Holland.

Vopěnka, P. (1979). *Mathematics in the alternative set theory*. Leipzig: Teubner.

Vopěnka, P., & Hájek, P. (1973). Existence of a generalized semantic model of Gödel-Bernays set theory. *Bull. Acad. Polon. Sci., Sér. Sci. Math. Astronom. Phys.*, *XXI*(12).

Vítězslav Švejdar
Department of Logic, Charles Univ. in Prague
Palachovo nám. 2, 116 38 Praha 1
e-mail: vitezslav.svejdar@cuni.cz
URL: http://www1.cuni.cz/~svejdar/

Prospects for a Comparative Analysis of Set Theory and Semantics

Giulia Terzian[*]

Abstract

There is a structural similarity between the paradoxical phenomena to which the naïve versions of set theory and semantics are vulnerable, in virtue of satisfying a set of jointly inconsistent conditions. But can claims about paradoxes give a reliable underpinning to claims about their overarching theories? The paper gives a negative answer to this question and proposes a methodologically more sound route to a comparative analysis of set theory and semantics.

1 Introduction

Alfred Tarski's work on truth and the semantic paradoxes (chiefly Tarski, 1936) was ground-breaking in many respects, and contributed greatly to shape the modern landscape of philosophical logic. One way to understand the great importance of Tarski's work is the following: Tarski was the first to invoke, and implement, *formal* constraints and desiderata for a philosophical account of the concept of truth. That is, he sought to produce a rigorous formal theory that preserved as many as possible of the essential features of natural language and of our intuitions about truth, which could then be studied from the logician's perspective, much like any other formal theory.

The core of Tarski's theory can be broken down to the conjunction of three formal constraints:

---

[*]This work is supported by the Arts and Humanities Research Council, UK.

(i) Self-referential expressions can be constructed in $\mathcal{L}$ (the base language)

(ii) Classical logic holds

(iii) For every $\varphi \in \mathcal{L}$, for Tr a unary predicate not in $\mathcal{L}$, the theory proves all instances of the truth scheme

$$Tr(\ulcorner \varphi \urcorner) \leftrightarrow \varphi$$

Satisfaction of (i)–(iii) make the theory for $\mathcal{L}_{Tr}$ vulnerable to paradox, since it becomes possible to prove sentences such as $\lambda \leftrightarrow \neg Tr(\ulcorner \lambda \urcorner)$ (the liar sentence), from which in turn the contradiction $\lambda \leftrightarrow \neg \lambda$ can be derived.

Keeping in line with most of the literature on the topic, in this paper liars are taken as representatives of all semantic antinomies (this is possible again thanks to Tarski, who showed that all semantic notions can be reduced to a common denominator, namely the satisfaction of a formula by a sequence of objects[1]).

This latter umbrella-like categorization of the semantic paradoxes has sometimes been taken more or less at face value: F. P. Ramsey famously argued that the paradoxes involving semantic concepts should be separated from those involving "only logical or mathematical terms such as class and number" (Ramsey, 1931, p. 353), i.e. pertaining to the domain of set theory; and accordingly, the respective disciplines should also be kept apart.

Today, most philosophers reject Ramsey's classification. Perhaps the main complaint is that it neglects the fact that certain patterns would appear to exist and to be shared by both set-theoretic and truth-theoretic (or semantic) paradoxes. An argument to this effect is put forth notably by Priest (1994), who presses the conclusion that we should look for one solution path rather than two (or more). On the basis of a technical analysis of some of the main antinomies, Priest champions a 'principle of uniform solution' for paradoxes of semantics and set theory alike. (For earlier examples of similar lines of argument see e.g. Herzberger, 1970 and Feferman, 1984.)

In the next section Priest's claim is critically examined.

---

[1] See e.g. (Tarski, 1936).

2 The analogy argument

Some formal preliminaries:

It is assumed throughout the paper that the theories being discussed are formulated in a first-order language $\mathcal{L}$ with identity, whose syntax is entirely codifiable in the language of arithmetic $\mathcal{L}_{arith}$. The standard notation of $\ulcorner\varphi\urcorner$ is used to denote the gödel code of φ.

Then $\mathcal{L}$ satisfies condition (i) above.

Two extensions of the base language are considered: the first with the predicate symbol Tr, the second with the relation symbol $\in$. Then condition (iii) above will be satisfied if the unrestricted comprehension and truth schemes are assumed to hold in the theories for the languages $\mathcal{L}_{Tr}$ and $\mathcal{L}_{\in}$ respectively:

(CA) $\exists x \, \forall y \, (y \in x \leftrightarrow \varphi[y])$

(TA) $Tr(\ulcorner\varphi\urcorner) \leftrightarrow \varphi$.

The resulting theories, in other words, are the inconsistent (naïve) versions of set theory and formal semantics, respectively.

Up to this point the setup was of course intended to highlight the analogies between the set-theoretic and the semantic contexts. It is just the presence of such parallel features that seems to have motivated Priest to advocate the claim that the paradoxes ought to have the same solution.

What is meant by a "solution" to the paradoxes? In the set-theoretic case, the answer is simple: the ZF axioms, i.e. the most successful theory of sets to date. What about the semantic case? Here the answer cannot be equally punctual, because the semantic paradoxes have not been "solved" to widespread and shared satisfaction.

We have seen that Priest moves from the claim that the paradoxes have a similar underlying mechanism to the claim that a corresponding similarity must exist between the respective solutions. I claim the analogy argument proposed by Priest is methodologically unsound.

The notion of solution to the (set-theoretic or semantic) antinomies that is invoked in the literature is typically a broad one, amounting to more than a simple tweak in the formal apparatus of the theory devised uniquely to block out the inconsistencies. Both set theorists and philosophers generally think of solutions to the paradoxes as being subsumed within, and being immediate consequences of, the formal explication of the central concepts of the respective domains of discourse; in other words, the respective theories.

Liar-like sentences are extreme, even degenerate, instances of application of the concept of truth; just like non-self-membership is a degenerate instance of application of the concept of set: this much is common ground. But in virtue of this fact, the paradoxes should not be not points of departure, but rather test-beds, for the theory in question.

The methodological objection can now be pinned down more concisely: analogies in the derivation of the paradoxes cannot constitute sufficient evidence that some (the same) analogy is also found at the level of the theories.

Still, the point of *dis*analogy between set theory and semantics—a solution is available in the first case only—justifies interest in the project of their comparative analysis. The rest of the paper attempts to investigate whether the analogy claim is tenable.

I begin by suggesting that in order to secure methodological defeasibility the analogy argument should have the following form:

(Par) Set-theoretic and semantic paradoxes have the same structure.

(Th) The theories of sets and truth have the same structure.

(ZF) ZF "solves" the set-theoretic paradoxes.

∴

(Sol) Any solution to the semantic paradoxes must (is expected to) have the same structure as ZF set theory.

This is not yet an operative argument, because premise (Th) is quite vague; I consider two possible ways of fleshing it out:

a. (Norm) The theory of sets and the theory of truth follow the same *underlying norms*.

b. (Ref) The theory of sets and the theory of truth prove *reflection principles* of the same form.

The rest of this paper will be devoted to assessing the soundness of the analogy arguments obtained by replacing (Th) in turn with (Norm) and (Ref).

3 Norms and the analogy

The first suggestion is to test the analogy account starting from the very bottom: if some genuine, interesting, substantial analogy can be made between set theory and semantics, then this should already show up at the lowest and most basic level of the theories, i.e. their axiomatic foundations.

3.1 The set-theoretic side of things

The literature points to two principles as the key underlying constraints of the ZF axioms. These are the *limitation of size conception* and the *iterative conception*:

(IT) A collection of objects is a set iff it is produced in a *linear, well-founded* process, according to the following inductive definition:

$$V_0 = \emptyset; \qquad V_{\alpha+1} = \wp(V_\alpha); \qquad V_\gamma = \bigcup_{\beta<\gamma} V_\beta \text{ for limit } \gamma.$$

(LIM) A collection of objects is a set iff it is *small enough*: that is, iff the objects of the collection are not in one-to-one correspondence with the universe of sets.

The goal of this section can now be reformulated more precisely as that of evaluating whether any relevant semantic principles can be found to correspond to (IT) and (LIM). Equally importantly, the semantic principles should be foundational principles: they ought to embody (some of) the most fundamental intuitions about the concept of truth of which the theory provides a formal explication.

3.2 Iterative conception, hierarchies, dependence

The principle (IT) describes the process by which sets "come into existence"; or to use a more neutral expression, it describes the inductive procedure underlying the construction of the set-theoretic universe. The fundamental features are (i) iteration: the sets appearing at a given level are collected into a new set, situated at either the next successor or the next limit stage; (ii) hierarchy: each stages of the inductive definition is more inclusive than the previous; (iii) cumulative: a set can only enter the hierarchy, never exit.

Something else that emerges from (IT) is that sets, and indeed the hierarchy, are intrinsically characterized by the obtaining of a *dependence* relation. Any set *depends* upon its members; something which is sometimes conveyed by saying that there is nothing more to a set than its members. The inductive definition of the V-hierarchy expresses the fact that each set is an end-point of a dependence path that goes all the way down to the origin of the hierarchy.

Furthermore each level of V *depends* upon the previous level(s). The dependence relation itself is embodied by the membership sign $\in$; and the latter is in turn defined implicitly by the ZF axioms. So $\in$-dependence (as it will be referred to hereafter) is at the heart of the set-theoretic hierarchy, and conversely the axioms provide a full characterization of $\in$-dependence.

Kripke's inductive definition of a partial truth predicate for a first-order language gives rise to a construction which shares certain fundamental features with the set-theoretic hierarchy. In both cases, the objects of the domain of quantification—sets, sentences—make an appearance at some determinate stage of the inductive definition; once a particular element enters the hierarchy, it is always present at later stages; finally, from any level σ, σ-many iterations of the inverse power-set and jump operations (respectively) eventually lead back to stage 0 of the construction.

Kripke's choice of a partial truth predicate led him to make use of the non-classical logic scheme of Strong Kleene; in this respect the semantic theory is starkly different from the set-theoretic. However, perhaps this is not as problematic a difference as may appear at first; for instance, the compositional notion of dependence is also invoked by Yablo (1982), who makes use of a supervaluation scheme and thus reverts to a bivalent logic.

A more important difference lies in the fact that $\in$-dependence but not K-dependence underlies criteria of identity for its relata. Considering a concrete case will help clarify this claim. Take the liar sentence $\lambda \leftrightarrow \neg Tr(\ulcorner \lambda \urcorner)$: the Kripkean or compositional analysis of its dependence history tells us that whether or not $\neg Tr(\ulcorner \lambda \urcorner)$ is placed inside the extension of Tr directly depends on whether λ is. Since λ is materially equivalent to, and therefore shares its dependence path with, $\neg Tr(\ulcorner \lambda \urcorner)$, the familiar circular pattern arises; however, nothing in the compositional analysis gives a criterion to discriminate such circular dependence paths from the well-founded.

Compositional dependence, or K-dependence, does not exhaust the application of this concept in semantics. An alternative account is developed by Leitgeb (2005). In brief, the aim in Leitgeb's paper is to identify a class $\mathcal{L} \subset \Phi \subset \mathcal{L}_{Tr}$ consisting of all and only those sentences $\phi \in \mathcal{L}_{Tr}$ that *depend* directly or indirectly on non-semantic states of affairs.

Formally, this is accomplished by defining a dependence operator $D(\Phi)$ which is monotone inductive and therefore has a least fixed point Φ_{lf}; the least fixed point has the attractive feature of containing the unrestricted instances of all of its T-biconditionals: $\forall \varphi \in \Phi_{lf} (Tr(\ulcorner \varphi \urcorner) \leftrightarrow \varphi)$.

The condition for a sentence φ to depend on another, or more generally on a set of sentences $\Phi \subseteq \mathcal{L}_{Tr}$ is thus as follows:

$$\varphi \text{ depends on } \Phi \text{ iff } \forall \Psi \subseteq \mathcal{L}_{Tr} : Val_{\Psi}(\varphi) = Val_{\Psi \cap \Phi}(\varphi)$$

The first observation is that Leitgeb's notion of dependence is not specifically tied to the truth-theoretic context: "Since Tr is left uninterpreted and is a fortiori not yet regarded as expressing truth, our theory of dependence is not specifically a part of a theory of truth but its scope is more general; other applications of the theory are conceivable" (Leitgeb, 2005, p. 174); for instance, to set theory.

Technically, the definition of the dependence operator as a monotone inductive operator has the consequence that the resulting theory is structurally comparable to the set-theoretic hierarchy. In Leitgeb's own words: "if the domain of this dependence relation is restricted to the set of sentences that depend on non-semantic states of affairs, a well-founded hierarchy of dependency up to the ordinal level of [the least fixed point] is determined. This is [...] analogous to the situation in set theory, where [...] if the domain of the membership relation is restricted to the class of well-founded sets, membership turns out to be arranged according to a similar ordinal system of levels (though, of course, of "unbounded height")." (Leitgeb, 2005, pp.174–175)

The choice of a supervaluational scheme ensures moreover that classical logic underlies Leitgeb's account, which thus resembles ZF set theory in this respect also.

Concerning the definition of L-dependence more specifically, it can be noted that "[this] notion of dependence [...] is a kind of *supervenience*" (Leitgeb, 2005, p. 160). In other words, the truth value of φ cannot change without change in the extension of the truth predicate.

One way of understanding the nature of this "kind of supervenience" is then to think that the obtaining of an instance of the dependence relation requires that the semantic facts about φ and ψ, *and* the extension of Tr, are already settled. Contrast this with K-dependence: on the compositional analysis of dependence the process of semantic evaluation is articulated in stages that mirror the syntactical complexity of the sentences. Instead one of the ideas behind Leitgeb's account is that over and above syntactical complexity, the inductive process by which truth values are assigned to sentences of $\mathcal{L}_{Tr}$ should take into account the number of iterated truth predications in φ. Thus at each stage there will be a collection of sentences that are fed into the evaluating mechanism together, rather than one at a time; and the results of the evaluation at each stage—the adding of new sentences to the extension of Tr—will go to determine which sentences will be placed inside that extension at the next stage.

To summarise: Sentences, it has been seen, can stand in two kinds of dependence relation: the more direct relation akin to compositionality, namely K-dependence; and the technically and conceptually more complex notion akin to supervenience, namely L-dependence. Compositional dependence gives no normative grip to settle questions about truth-evaluability; L-dependence gives a way of settling such questions.

Returning to set theory, it certainly seems as though (IT) has normative force: a collection only qualifies as a set if it is the output of the iteration process; and conversely, any collection violating (IT) has no place in the universe of sets.

It may be concluded that L-dependence stands as the best candidate for a semantic counterpart to (IT).

3.3   Limitation of size: size of what?

The iterative conception gives a criterion for distinguishing sets from non-sets, by specifying the "right" construction procedure. But (IT) on its own does not in fact suffice as a demarcating criterion for the admissible sets. A formal proof of this fact is given by Boolos (1998), who shows that (IT) and (LIM) are individually necessary and *jointly* sufficient to generate the ZF axioms.

The principle (LIM) rules out from qualifying as sets those collections of cardinality greater than, or equal to, the cardinality of the entire set-theoretic universe. For if it were possible to construct a set that matched the universe in size, then one more iteration of the power-set operation would produce a set strictly larger than the previous; so that the resulting set—formed in accordance to the iterative conception—would either fall outside of the universe of sets, or it would fall within but would be larger than its superset. Both options lead to contradiction; this is of course by and large the argument known as Cantor's paradox. Moreover, choosing instead to forbid the last iteration of the power-set operation would clearly contravene the iterative conception.

In a similar spirit to the previous section, we now look for a semantic counterpart to (LIM). Thus we are led to look for a principle that places an upper limit on the size of the arguments of the truth predicate; that is, the sentences of $\mathcal{L}_{Tr}$. But what is the "size" of a sentence?

A first possibility is to say that what determines the size of a sentence is the size of its predicate-extensions.

Here is a candidate for a semantic limitation of size principle:

(LIM$_{Tr}$) A sentence of $\mathcal{L}_{Tr}$ expresses a proposition iff it is *small enough*: iff its predicate-extensions are smaller in cardinality than $\mathcal{L}_{Tr}$.

We now examine three test cases for the new principle.

Consider the sentence $\phi \equiv \forall x(P(x) \rightarrow Tr(x))$, for some non-vacuous predicate P of $\mathcal{L}_{Tr}$. This is certainly a sentence that ought to be derivable from the theory of truth for $\mathcal{L}_{Tr}$.

No restrictions on which objects P is allowed to range over means no restriction on which objects Tr is allowed to range over; but if $\text{Ext}(P) = \text{Ext}(Tr) = \mathcal{L}_{Tr}$, then one possible instantiation of P, and therefore of *Tr*, is λ. So on this choice of extension ϕ fails the size test.

Suppose now that the extension of P is $\mathcal{L}_{arith}$. Arithmetical statements never feature the truth predicate, so no conflict should arise with (LIM$_{Tr}$). However $\mathcal{L}_{arith}$ and $\mathcal{L}_{Tr}$ have the same cardinality: thus the size principle gives two contradictory verdicts where it would be expected to give the same.

An even more conclusive example: take the extension of P to be the single sentence λ. There is no doubt that if the assertability of our sentence were to depend on the threshold for acceptability of *Tr*-extensions, the choice of $\{\lambda\}$ as such an extension should be forbidden; but it seems incoherent to appeal to cardinality considerations in relation to a singleton set. Another strike for the analogy account.

Here is an alternative suggestion: take *quantifier ranges* to be the objects of the (cardinality-like) constraint. This leads immediately to look in the direction of the debate over absolutely general quantification. Space constraints prevent from exploring further this direction of the analogy; the conjecture is that it would prove interesting, even if not ultimately fruitful for the analogy account, to do so.

4   Reflection and the analogy

In this final section the attention is turned to the meta-theoretic resources of ZF set theory, as another possible direction of comparison with semantics.

Reflection principles are well known, very fruitful meta-theoretic principles in set theory. Roughly speaking a typical reflection principle says that for any given statement A that holds in the universe of sets, there is some ordinal level α of the V-hierarchy at which A^α—A relativized to V_α—holds. In other words, any set-theoretic claim that is true in V, or true simpliciter, has a witness in the hierarchy.

Formally, and for all $\varphi \in \mathcal{L}_{Set}$:

(RP) $\forall\beta \, \exists\alpha > \beta \, \forall x \, (\phi(x) \leftrightarrow \phi^{V_\alpha}(x))$

The form of reflection principles is that of soundness claims: any formula ϕ that is an axiom or a theorem of the theory can also be shown to be true in the theory.

Therefore reflection principles also serve to carry out consistency proofs of ZF theory. Gödel's Theorems show that the consistency statement Con(ZF) cannot be derived in ZF; but it can be derived in the augmented theory ZF + Ref(ZF); moreover this ascent to stronger theories can be iterated. Moreover, one can have reflection principles for extensions of ZF theory with large cardinal axioms, which can also be proved sound, and therefore consistent, by the same method. Finally, reflection principles are also used to justify some of the ZF axioms themselves; for instance, (RP) "turns out to be equivalent to the two axioms infinity and replacement together, and ti can be regarded as giving a useful insight into what these two axioms say about the cumulative type structure" (Drake, 1974, p. 98).

Since reflection principles have such an important role in set theory, indeed so many connected important roles, the question arises of whether structurally analogous reflection principles could or do play analogous meta-theoretical roles for theories of truth.

On the basis of the features of set-theoretic reflection principles noted so far, it is to be expected that truth-theoretic reflection principles would be of the form:

$$\mathbf{T}_{Tr} \models Prov(\ulcorner \phi \urcorner) \rightarrow \mathbf{T}_{Tr} \models Tr(\ulcorner \phi \urcorner)$$

Or more concisely

$$\mathbf{T}_{Tr} \models Prov(\ulcorner \phi \urcorner) \rightarrow \phi$$

Indeed the reflection principles discussed in the literature on truth (see e.g. Ketland, 1999) have precisely this form of soundness principles and thus match the expectation. Although again space constraints prevent from developing these considerations into a solid argument, it appears at least that the analogy account gains further credibility.

5 Conclusions

This paper addressed the question of whether a structural analogy between set theory and semantics can plausibly be carried out. The investigation of the analogy account has delivered mixed results; hopefully one of its upshots is to have brought methodological transparency to this important and interesting issue.

References

Boolos, G. (1998). Iteration again. In *Logic, logic and logic*. Harvard University Press.

Drake, F. R. (1974). *Set theory: an introduction to large cardinals*. North-Holland.

Feferman, S. (1984). Toward useful type-free theories. I. *Journal of Symbolic Logic, 49*(1), 75–111.

Herzberger, H. G. (1970). Paradoxes of grounding in semantics. *Journal of Philosophy, 67*(6), 145–167.

Ketland, J. (1999). Deflationism and Tarski's paradise. *Mind, 108*(429), 69–94.

Kripke, S. A. (1975). Outline of a theory of truth. *Journal of Philosophy, 72*(19), 690–716.

Leitgeb, H. (2005). What truth depends on. *Journal of Philosophical Logic, 34*(2).

Parsons, C. (1974). The liar paradox. *Journal of Philosophical Logic, 3*, 381–412.

Priest, G. (1994). The structure of the paradoxes of self-reference. *Mind, 103*, 25–34.

Ramsey, F. (1931). The foundations of mathematics. In Braithwaite (Ed.), *The foundations of mathematics and other logical essays, by Frank Plumpton Ramsey*. London: London: Routledge Kegan Paul.

Tarski, A. (1936). The concept of truth in formalized languages. In *Tarski (1956), pp. 152–278*.

Tarski, A. (1944). The semantic concept of truth and the foundations of semantics. *Philosophy and Phenomenological Research, 4*(3), 341–376.

Tarski, A. (1956). *Logic, semantics, metamathematics* (J. Woodger, Trans.). Oxford: Clarendon Press.

Yablo, S. (1982). Grounding, dependence, and paradox. *Journal of Philosophical Logic, 11*(1).

Giulia Terzian
Department of Philosophy, University of Bristol
9 Woodland Road, Bristol BS8 1TB, UK
e-mail: `giulia.terzian@bristol.ac.uk`